Atomic Diffusion Theory
in
Highly Defective Solids

Diffusion and Defect Monograph Series

Edited by:

Y. Adda, A. D. Le Claire, L. M. Slifkin, and F. H. Wöhlbier

Advisory Board:

B. I. Boltaks, G. Brebec, Th. Heumann, K. I. Hirano,
H. B. Huntington, D. Lazarus, A. Lodding, A. J. Mortlock, J. Philibert,
R. Reisfeld, S. J. Rothmann, R. A. Swalin, and H. Wever

No. 1 (1972)

Surface Self-Diffusion of Metals

G. Neumann and G. M. Neumann

No. 2 (1973)

Electrotransport in Metals and Alloys

J. N. Pratt and R. G. R. Sellors

No. 3/4 (1975)

Ionic Diffusion in Oxide Glasses

G. H. Frischat

No. 5 (1977)

Diffusional Creep of Polycrystalline Materials

B. Burton

No. 6 (1980)

**Atomic Diffusion Theory
In Highly Defective Solids**

G. E. Murch

Atomic Diffusion Theory
in
Highly Defective Solids

G. E. Murch

Argonne National Laboratory
Chemistry Division
Argonne, Ill. 60439, USA

Trans Tech SA
Trans Tech House
4711 Aedermannsdorf, Switzerland

Copyright © Trans Tech SA Switzerland

ISBN 0–87849–516–9

Distributed in North America by:

Trans Tech Publications c/o Karl Distributors
16 Bear Skin Neck
Rockport, Mass. 01966
USA

and worldwide by

Trans Tech SA
Trans Tech House
4711 Aedermannsdorf
Switzerland

Printed in USA

Table of Contents

TABLE OF CONTENTS

1. Introduction

About a decade ago it became clear that highly defective
solids, as a group, were not covered within the precincts of
random walk diffusion theory. These solids introduced special
difficulties which could not be handled by any straightforward
extension of a theory which was developed primarily for metals
and stoichiometric binaries.

The principle difficulties may be stated quite simply.
Both implicitly and explicitly, diffusion theory had been concerned
with solids with very low defect concentrations. Immediately
then, highly defective solids introduced a new variable, defect
concentration, to such quantities as the tracer correlation
factor. More importantly, however, as a result of a high concentration
of defects, unlikely coordinations of some atoms would occur
with a purely random distribution of defects. Thus some degree
of partial ordering of defects through defect-defect or, more
correctly, atom-atom interactions must naturally be assumed. The
traditional distinction between an independent and mobile defect
species and a structural element now becomes obscured as a range
of defect mobility emerges. In effect, any previous equivalence
between mobile defect concentration and composition is now entirely
removed. Another aspect of this defect distribution problem
is that the nature of the defect itself can become quite complicated,
most likely as a result of the optimization of local interatomic
distances. Possibly such clusters enjoy an independent existence
as lone defects which are capable of directly providing a vehicle

for diffusion. Alternatively, their participation in diffusion
may be indirect, such as providing sources and sinks for some
more mobile species such as point defects or smaller cluster
fragments.

Within the last ten years a quantitative theory to deal
with such situations has been developed. The theory is, for
the most part, an extension of statistical mechanics into the
time-dependent regime. Despite this origin, the theory is expressible
ultimately in terms of traditional diffusional parameters and
this aspect has ensured a consistent mesh with random walk theory.
The theory has primarily been developed with the aid of two
powerful techniques: the first, the Path Probability method; the
other, the Monte Carlo method.

These two techniques, as applied to highly defective solids,
have heralded a number of exciting new concepts in solid state
diffusion theory. One of these is the discovery of a tracer
correlation factor greater than unity. This result, which arose
from a diffusion mechanism postulated for a defect cluster, was
the first of its kind in solid state diffusion theory. Another
discovery is the existence of a new correlation factor, the *physical*
correlation factor, which owes its origin to the non-random motion
of *untagged* atoms as induced by local order, obstacles or site
preference. Inclusion of this factor leads to a rigorous form
of the Darken equation appropriate to highly defective solids.

At the end of ten years of sustained growth, the subject
has now matured to the extent that there is general agreement

on its scope, content and its likely course of development. It
seemed appropriate to review the progress that had been made
and to disseminate this information in book form. The present
monograph was the result of such an intention. It was written
in such a way as to provide a *first* specialist text. Throughout
the monograph we have placed emphasis on an *understanding* as
provided by statistical mechanical arguments. We have assumed
that the reader has been exposed to some statistical mechanics
as found in an introductory text, for example, Hill (1962). We
have also assumed some exposure to diffusion theory; for this
we recommend the introductory text by Shewmon (1963) and the
extensive and general treatise by Adda and Philibert (1966).
For a specific background in tracer correlation effects we recommend
the text by Manning (1968) and the extensive review by LeClaire
(1970).

We have been somewhat arbitrary in our definition of what
we consider to be a highly defective solid. We have chosen
to focus on solids which *apparently* exhibit a high concentration
of mobile defects. Thus those *stoichiometric* compounds, in
which the structural interstices are significantly occupied
at high temperatures, qualify as being highly defective in our
definition. Many superionic conductors fall into this category.
The other more traditional group of highly defective solids
includes what are generally classed as 'nonstoichiometric' compounds.
This category includes, for example, the large group of transition
metal oxides, interstitial solid solutions and intercalation
compounds.

The organization of the monograph is as follows. In Chapter 2 we have presented a discussion of statistical thermodynamics with emphasis on the structure-sensitive partial molar thermo-dynamic quantities. This chapter is intended as a prelude to Chapter 3 where we have presented an exposition of atomic transport with an emphasis on a statistical mechanical approach. Within this chapter we have devoted a special section (3.5) to the detailed description of the Path Probability and Monte Carlo techniques. Results of the application of these techniques have already provided the basis of much of the understanding of kinetic processes occurring in highly defective solids. Further applications would seem to be useful and rewarding.

In neither of these chapters have we attempted to review the experimental data, although we have not hesitated to use experimental data to illustrate theoretical principles. We have, however, devoted some space, Appendix II, to a discussion of some recent experimental techniques in tracer diffusion which, in our opinion, have made, or are likely to make a significant impact on the acquisition of diffusion data in highly defective solids.

2. Statistical Thermodynamics

2.1 Introductory Remarks

In a discussion of statistical thermodynamics we have found it convenient to concentrate on the partial molar quantities and, in particular, their compositional dependence. Accordingly, we have restricted our discussion in this section to the considerable progress which has been made on the understanding of those highly defective solids which exhibit a compositional variability, i.e., nonstoichiometric compounds.

One may define a nonstoichiometric compound in either thermodynamic or structural terms. In a *thermodynamic* sense, at a given temperature and within a single phase region, the chemical potentials of both components are *continuous* functions of composition. Many systems are now known, however, where ordered phases with extremely narrow ranges of homogeneity exist at low temperature and which apparently smear out at higher temperatures to a single phase with a wide range of homogeneity. The thermodynamic properties of such systems have been examined under equilibrium conditions and the existence of the wide range nonstoichiometric phase is still assured. In a *structural* sense nonstoichiometric compounds can be characterized by diffraction means as being single phase and which may retain the residues of the crystallographic symmetry of the low temperature line phases(s).

Statistical thermodynamics provides the formal link between the structural and thermodynamic bases of nonstoichiometry. In principle, a satisfactory theory of nonstoichiometry should be consistent with the structural and thermodynamic definitions above. No such complete theory has yet been devised. Significant progress has, however, been achieved along essentially two lines of approach. In the first, nonstoichiometry is described

on the basis of an essentially disordered distribution of point
or clustered defects. In the second, one starts with an
essentially ordered state and introduces disorder as a pertur-
bation. When these points of view converge in their descriptions
of the nonstoichiometric phase one probably will have sufficiently
correlated the structural and thermodynamic definitions of
nonstoichiometry.

We now discuss the first point of view and it is convenient
to do this by tracing its history.

2.2. Approach from the Disordered State

2.2.1 Non-Interacting Point Defects

2.2.1.1 One Type of Defect

The statistical thermodynamics of a solution of point defects
of a single type which do not interact is the simplest possible
case to consider. While the assumption of non-interaction
clearly cannot be valid for any real system, consideration of a
single type of defect is appropriate to interstitial solid
solutions,intercalation compounds and those nonstoichiometric
compounds sufficiently far from stoichiometry that the influence
of the complementary defect of, say, the Frenkel pair is
negligible. Exemplifying the discussion with interstitials we
write for the petit canonical partition function, Q

$$Q = \frac{1}{N_i!} \sum_{j(states)} \exp(-E_j/kT), \qquad (2.1)$$

where N_i is the number of interstitials distributed over
γB sites and γ is the ratio of interstices to regular lattice sites,
E_j is the energy of state j and the summation is over all states
but with the specific condition that no more than one interstitial
occupies a site. The configurational energy for *every* state is

$$E_j = N_i E_i,$$ (2.2)

where E_i is the formation energy of an isolated interstitial.
The free energy, F, is defined by

$$F = -kT\ln Q,$$ (2.3)

and the chemical potential, μ_i, by

$$\mu_i = \left(\frac{\partial F}{\partial N_i}\right)_{B,T},$$ (2.4)

Using the usual maximum-term method (Hill 1962) we find that

$$\mu_i = -kT\ln\left(\frac{1-\Theta_i}{\Theta_i}\right) + E_i,$$ (2.5)

where $\Theta_i = N_i/\gamma B$.

Thus the entropy of solution is described as ideal while the heat
of the solution is constant. Historically, Eqn 2.5 is the three
dimensional analogue of the Langmuir adsorption isotherm.

2.2.1.2 Two Types of Defect

In order to traverse the homogeneity range of a nonstoichiometric compound between two diphasic regions, one requires two defect types, e.g., an anion vacancy and an interstitial. An example is $UO_{2\pm x}$ which can exhibit quite a wide homogeneity range between U/UO_{2-x} and UO_{2+x}/U_4O_9. The first treatment of the problem (Schottky and Wagner 1930) is not capable of including phase separation because there are no defect interaction terms (see section 2.2.2). In essence, their treatment was simply an extension of eqn 2.1 to include another defect type. Assuming Frenkel defects they wrote the equivalent of the following petit canonical partition function

$$Q = \frac{1}{N_i! N_v!} \sum_j \exp(-E_j/kT), \qquad (2.6)$$

with $E_j = N_v E_v + N_i E_i,$ \qquad (2.7)

and N_v is the number of vacancies and E_v is the energy to form an isolated vacancy. Upon using the maximum-term method and taking partial derivatives with respect to N_i and N_v we obtain the following expressions for the chemical potentials N_i and N_v

$$\mu_i = \left(\frac{\partial F}{\partial N_i}\right)_{N_v, B, T} = -kT\ln\left(\frac{1-\theta_i}{\theta_i}\right) + E_i \qquad (2.8a)$$

$$\mu_v = \left(\frac{\partial F}{\partial N_v}\right)_{N_i, B, T} = -kT\ln\left(\frac{1-\theta_v}{\theta_v}\right) + E_v \qquad (2.8b)$$

Let us now define, C_i, the intrinsic disorder by

$$C_i = \gamma \exp [-(E_v+E_i)/2kT].\qquad(2.9)$$

δ, the deviation from stoichiometry, is given by

$$\delta = (N_v - N_i)/B .\qquad(2.10)$$

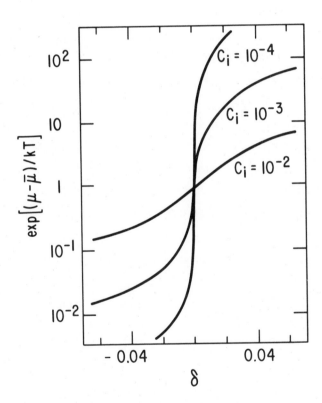

Fig. 2.1.   The dependence of $\exp [(\mu-\bar{\mu})/kT]$ on deviation from stoichiometry δ at several values of the degree of intrinsic disorder C_i.

If we define $\bar{\mu}$ as the chemical potential at the stoichiometric composition then it follows that the composite chemical potential μ is

$$\exp\left[(\mu-\bar{\mu})/kT\right] = \frac{-\delta + \sqrt{\delta^2 + 4C_i^{\ 2}}}{2C_i} \quad , \tag{2.11}$$

and this function is plotted in fig 2.1 for several values of C_i. It can immediately be seen that the degree of nonstoichiometry depends directly upon the level of intrinsic disorder of the stoichiometric crystal.

We have exemplified the above discussion with Frenkel type disorder. Anti-Frenkel and Schottky disorder yield a similar result.

Brebrick (1958) extended the above treatment to include ionization of defects. Thus a vacancy in the cation (or anion) sublattice was considered to be associated with an acceptor (or donor) level in the band structure. Each level was assumed to have a maximum occupancy of one and a degeneracy of two. The correlation between δ and the electronic band structure was described by the introduction or removal of the donor/acceptor levels. The compound is correspondingly an n or p-type semiconductor depending on whether electrons are added to the conduction band or holes are added to the valence band respectively.

We emphasize that non-interaction treatments such as these are only appropriate for compounds which exhibit very small deviations from stoichiometry and even then, the assumption of ideality is often highly questionable. More generally, one wishes to generate the phenomenon of phase separation as a consequence of the statistical treatment. This is only possible by introducing defect - defect interactions.

2.2.2 Interacting Defects

2.2.2.1 One Type of Defect

We return firstly to the case of a *single* type of defect.
The development of this field owes a great deal, at least in the
early years, to the quite analogous distribution problem of an
adsorbed species on a surface and to a lesser extent, order/disorder
in concentrated alloys. We exemplify the discussion again with
interstitials and we assume pairwise nearest neighbour interactions. One still retains the petit canonical partition function
(eqn 2.1) but the energy of each state is now given by

$$E_j = N_i E_i + N_{ii}^j E_{ii} , \qquad (2.12)$$

where N_{ii}^j is the number of nearest-neighbour interstitials in
state j and E_{ii} is the nearest-neighbour interaction energy.
The magnitude of E_{ii} and, indeed, sometimes even the sign are
not usually known *a priori*. In most cases it is treated as an
adjustable parameter when fitting the calculated chemical
potential to the experimentally determined chemical potential.
In some cases it can be calculated from the critical temperature
(see below). For many solutions of hydrogen in metals, E_{ii} is
thought to be < 0 (attraction) because of the reduction in local
strain which comes from the clustering of hydrogen atoms. For
oxygen, carbon and nitrogen in metals, E_{ii} is thought to be > 0
(repulsion) because those interstitials would be expected to
distort the metal lattice to the extent that the occupation of
neighbouring interstices is made less likely. For nonstoichiometric
compounds, many of which are quite ionic, e.g., UO_{2+x}, the
interaction of the defects is undoubtedly coulombic, at least at
long range. In this case, repulsion between nearest neighbour

defects can be considered perhaps a crude approximation to this
or an approximation for strain and overlap repulsion at close
distances.

The problem may be solved at various levels of approximation
but, as yet, not exactly in two or three dimensions except via
Monte Carlo calculations. At the zeroth or Bragg-Williams
approximation the defects are considered to be distributed
quite at random. The ensemble average configurational energy,
$<E>$, is then evaluated on this basis. $<E>$ is firstly written
in the general form

$$<E> = N_i E_i + <N_{ii}> E_{ii},$$ (2.13)

where $<N_{ii}>$ is the most probable number of interstitial pairs
and is given at the level of a random distribution by

$$<N_{ii}> = \frac{ZN_i^2}{2\gamma B}.$$ (2.14)

Again upon using the maximum-term method and eqn 2.4 we find
that the chemical potential is given by

$$\mu_i = -kT\ln\left(\frac{1-\Theta_i}{\Theta_i}\right) + [E_i + Z\Theta_i E_{ii}].$$ (2.15)

This equation illustrates, of course, an ideal entropy and a heat
of solution given by the term in brackets [ ], which is linear
in composition. Eqn 2.15 generates isotherms of the form of fig.
2.2.

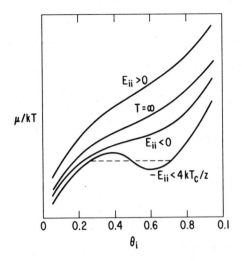

<u>Fig. 2.2.</u>   A sketch of the functional form of eqn 2.15 above and
below the critical temperature T_c.

When $E_{ii} < 0$, one finds the well-known condensation behaviour
into two co-existing phases, one a gas-like phase, the other a
liquid-like phase. The critical temperature at $\theta_i = 0.5$ is
given by

$$T_c = \frac{-ZE_{ii}}{4k} \; .$$
(2.16)

Below T_c we note the van der Waals loops. These unrealistic
loops are a consequence of the system being forced to be
homogeneous under all conditions; this is, of course, implied
in the assumption of a random distribution at the Bragg-Williams
approximation.

For repulsive interactions it can be seen from fig 2.2 that
the isotherms simply become steeper as E_{ii} increases, (T decreases).
In an exact treatment of this problem one would intersect an
ordered phase centered about $\theta_i = 0.5$ except in the 'triangular'

lattices where, for example, in f.c.c., three ordered phases at θ_i = 0.25, 0.5 and 0.75 of the type Cu_3Au, $CuAu$, $CuAu_3$ would in fact be generated. The ordered phases can only be generated in a Bragg-Williams treatment provided a long range order parameter and separate sub-lattices are explicitly introduced (Shockley 1938). Such a treatment has recently been applied to nonstoichiometric compounds (Tateno 1979).

A considerably better approximation than the B.W. is the Bethe approximation which is also equivalent to the Quasi-Chemical approximation. In this case, *pairs* of nearest neighbour sites are treated as being independent of each other. In the 'triangular' lattices e.g., f.c.c., the pair approximation is quite poor since two nearest neighbours of a site can themselves be nearest neighbours. For the f.c.c. lattice, the basic cluster should be at least a tetrahedron, but this appears not to have been used in calculations of the lattice gas chemical potential. A tetrahedron and higher clusters have been used in the investigation of order/disorder in the analogous binary alloy where an A atom corresponds to an interstitial and a B atom corresponds to a vacant site (Kikuchi and Sato 1974).

Firstly, we start by writing down the number of interstitial-interstitial pairs, N_{ii}

$$2N_{ii} = ZN_i - N_{iv},\qquad(2.17)$$

and the number of vacancy-vacancy pairs, N_{vv}

$$2N_{vv} = Z(N - N_i) - N_{iv}\qquad(2.18)$$

where N_{iv} is the number of interstitial-vacancy and vacancy-interstitial pairs. The above conditions (eqns 2.17 and 2.18) are such that only one independent number of pairs exists, N_{iv}. The petit canonical partition function (eqn 2.1) can be recast as

$$Q = \sum_{N_{iv}} g(N_i, B, N_{iv}, T) \exp [-E_{ii}N_{iv}/2kT], \quad (2.19)$$

where g is the number of configurations with exactly N_{iv} pairs of the type interstitial-vacancy (and vice versa). That is to say, there are g different ways of distributing N_i interstitials on γB sites to give N_{iv} pairs of type i - v (and v - i). The result for $g(N_i, B, N_{iv}, T)$, which we will not derive here, is:

$$g(N_i, B, N_{iv}, T) = \left(\frac{(\gamma B)!}{N_i!(N-N_i)!}\right)^{1-Z} \left(\frac{(ZB\gamma/2)!}{N_{ii}!N_{vv}![(N_{iv}/2)!]^2}\right). \quad (2.20)$$

Using the maximum-term method we find that the chemical potential is given by

$$\mu_i = \frac{ZkT}{2} \ln \left\{\frac{[(R - 1 + 2\theta_i)(1 - \theta_i)]}{(R + 1 - 2\theta_i)\theta_i}\right\} + kT\ln \left(\frac{\theta_i}{1-\theta_i}\right) \quad (2.21)$$

$$+ \frac{ZE_{ii}}{2} + E_i,$$

with

$$R = \{[1 - 4\theta_i(1 - \theta_i)][1 - \exp(-E_{ii}/kT)]\}^{\frac{1}{2}}. \quad (2.22)$$

The heat of solution E_i^m is given by:

$$E_i^m = E_i + \frac{ZE_{ii}}{2} - \frac{Z}{2R} E_{ii}(1-2\theta_i), \quad (2.23)$$

which is to be compared with the bracketed term in eqn 2.15. In the limit $E_{ii}/T \to 0$ then eqn 2.23 reduces to the linear eqn 2.15. Eqn 2.23 generates a sigmoidal shape for E_i^m which is, in fact, seen in many systems, e.g., UC_x 1.0 < x < 2.0 (Tetenbaum and Hunt 1971).

For the nearest neighbour interaction problem one should not need to venture further to higher approximations since it is now known that the agreement between μ of eqn 2.21 *above* the critical temperature and *exact* Monte Carlo determinations is within 0.5% (Murch and Thorn 1978d). For the f.c.c. lattice the situation with regard to the pair approximation is, however, rather special as pointed out above. Since the tetrahedron approximation is rather *unwieldy* other approximations have been postulated. Alex and McLellan (1971) used the method of Kirkwood expansions to determine μ. In this method it can be shown that the petit canonical partition function can be written in the form

$$Q = \frac{N_i!}{(N-N_i)! \, (\gamma B)!} \left\{ 1 + \alpha M_1 + \frac{\alpha^2}{2!} M_2 + \frac{\alpha^3}{3!} M_3 + \ldots \right\}, \qquad (2.24)$$

where $\alpha = -E_{ii}/kT$,

and $M_j = \langle N_{ii}^j \rangle$ is the *a priori* average of the j'th power of N_{ii}. The free energy is written as

$$F = -kT \ln Q = N_i E_i - kT \ln \left(\frac{N_i!}{(N-N_i)! \, (\gamma B)!} \right) - kT \left\{ \alpha \lambda_1 + \frac{\alpha^2 \lambda_2}{2!} + \frac{\alpha^3 \lambda_3}{3!} \right\} \qquad (2$$

where $\quad \lambda_1 = M_1$

$$\lambda_2 = M_2 - M_1^{\,2} \qquad\qquad (2.26)$$

$$\lambda_3 = M_3 - 3M_2 M_1 + 2M_1^{\,3}$$

etc.

Krivoglaz and Smirnov (1964) have described in detail the calculation of the M_i. Results for the λ_i are (Alex and McLellan 1971)

$$\lambda_1 = \gamma 6B\Theta_i^2$$

$$\lambda_2 = \gamma 6B\Theta_i^2 (1 - \Theta_i)^2 \tag{2.27}$$

$$\lambda_3 = \gamma 6B[\Theta_i^2 (1 - \Theta_i)^2 (1 - 2\Theta_i)^2 + 8\Theta_i^3 (1 - \Theta_i)^3]$$

Although the series converges very slowly, for small values of E_{ii} and low values of Θ_i, terms only up to λ_4 need to be considered. Application has been made to carbon dissolution in γ-Fe (Alex and McLellan 1971). Recent Monte Carlo simulation in the same system has revealed the essential validity of that calculation (Murch and Thorn 1979c).

Foster and Dooley (1977), in their study of austenite, have taken a quite different approach by establishing an expression for μ in terms of a dynamic balance between paired and unpaired interstitials. They wrote for the jump frequency for unpaired interstitials jumping to paired positions with an activation energy E_{act}°

$$\omega_1 = \nu \exp (-E_{act}^{\circ}/kT). \tag{2.28}$$

Conversely, the jump frequency for paired interstitials is

$$\dot{\omega}_2 = \frac{11}{12} \nu \exp [-(E_{act}^{\circ} - E_{ii}/kT], \tag{2.29}$$

where ν is the lattice vibrational frequency. Foster and Dooley were then able to show that one could expand the number of possible configurations W as

$$W = \frac{\gamma B (N_i - 1 - 12\beta) \dots [\gamma B - (N_i - 1)(1 + 12\beta)}{N_i!}, \tag{2.30}$$

where

$$\beta = 1 - \frac{12}{11} \exp (-E_{ii}/kT).$$

This led to the following equation for μ_i

$$\mu_i = E_{ii} \left\{ 1 - \left[\frac{1 - \Theta_i}{1 - \Theta_i [1 - 48 \exp (-E_{ii}/kT)]} \right]^{\frac{1}{2}} \right\}$$

(2.31)

$$+ \ln \left[\frac{\Theta_i}{1 - \Theta_i [2 + 12\beta]} \right].$$

Foster and Dooley found that a value for E_{ii} of about 1 kcal/mole[-1] fitted the carbon activity data for austenite quite well. However, Murch and Thorn (1979c) in a Monte Carlo calculation, revealed that E_{ii} should in fact be about 1.83 kcals.mole[-1] in agreement, incidentally, with Alex and McLellan (1971). The discrepancy perhaps can be traced back again to the special difficulty in the f.c.c. lattice where nearest neighbours can themselves be nearest neighbours.

As alluded to above, Murch and Thorn (1976, 1978d, 1979c) have performed several Monte Carlo calculations for the problem of calculating μ in the nearest neighbour interaction problem. There are two distinct means of achieving this. The first, almost trivial method, makes use of the grand canonical (μ, B, T) ensemble; a chemical potential is assigned and the concentration of particles, say interstitials, equilibrates to this through the use of an external particle source/sink. The second method

makes use of the petit canonical (N_i, B, T) ensemble. In
this case, one writes the derivative ($\partial F / \partial N_i$) in the following
finite-difference form

$$\mu_i = \left(\frac{\partial F}{\partial N_i}\right)_{B,T} = \lim_{N_i \to \infty} -kT\ln \left[Q(N_i + 1, B, T)/Q(N_i, B, T)\right], \qquad (2.32)$$

which can be developed to give

$$\mu_i = -kT\ln \left[\frac{B}{(N_i + 1)} < \exp (-\Delta E_m/kT) > \right], \qquad (2.33)$$

where $< >$ denote an ensemble average and ΔE_m is the change in
the energy of state m on addition of the $N_i + 1$'th particle.
In essence, the derivative, $\partial F/\partial N_i$, is taken by hypothetically
adding a new particle to the system and assessing the energy
change. For further details on the Monte Carlo method we
refer to section 3.5.3.

Developing in parallel with the above treatments concerned
with the statistics of *soft* interactions between defects have been
treatments based on site exclusion, 'blocking' or 'hard' interaction between defects. The basic assumption here is that the
observed deviation from ideal behavior arises *entirely* from the
configurational entropy. This implies, of course, that there is
no energy change when a new defect is added to the system. As
a consequence, the heat of solution is independent of defect
concentration.

One considers that a given defect (vacancy, interstitial, defect complex, etc) blocks or excludes an integral number of neighbouring sites. In the case of interstitial solid solutions the justification for blocking is based on the perhaps not unreasonable premise that there is local strain on the lattice such that occupation of nearest neighbour sites is impossible. In the case of anion vacancies in a nonstoichiometric compound like CeO_{2-x} the justification for blocking is based on the fact that an unreasonable cation coordination would be called for if there were only a *random* distribution of anion vacancies (Barker and Knop 1971). Moreover, in such an ionic solid, more extended site exclusion may be thought of as a crude form of coulombic repulsion since vacancies in CeO_{2-x} are undoubtedly ionized and so carry an effective charge.

The concept of blocking has also been extended to defect complexes or clusters as found in some nonstoichiometric compounds. Defect clusters, which are quite coherently meshed in the host matrix, are, in essence, a result of a local reconstruction to give optimum interionic distances. Not surprisingly, defect clusters tend to resemble fragments of adjacent phases. In fact, the adjacent phase may be built up from an ordered arrangement of clusters. Well-known examples are the Willis (1964) clusters in UO_{2+x} (fig 2.3) and the Koch-Cohen (1969) clusters in $Fe_{1-\delta}O$ (fig 2.4). For clusters, the case for site exclusion is stronger than for point defects because it is immediately clear that a number of configurations for

neighbouring defect clusters must automatically be blocked
because a given defect cluster occupies more than one lattice
site.

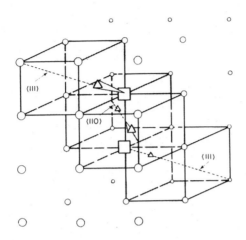

Fig. 2.3. The Willis (1964) 2:2:2. defect cluster in UO_{2+x}. The
cluster consists of 2 normal oxygen vacancies, 2 O' interstitials
in the <110> directions and 2 O" interstitials in the <111> directions.
The 2:1:2 cluster has only one O' interstitial.

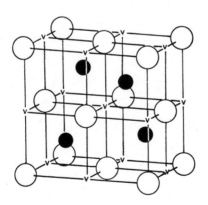

Fig. 2.4. The Koch-Cohen (1969) defect cluster in $Fe_{1-\delta}O$. The cluster
consists of 13 vacant cation sites (v) and 4 tetrahedral iron atoms ●.

Speiser and Spretnak (1955) were probably the first to
consider site blocking as a means of modifying the configurational
entropy. They wrote for the number of ways, W, of placing N_i
interstitials on B sites with each interstitial blocking α sites

$$W = \frac{\gamma B (\gamma B - \alpha)(\gamma B - 2\alpha) \cdots (\gamma B - [N_i - 1]\alpha)}{N_i !} \cdot \qquad (2.34)$$

The configurational free energy, E, is given by

$$F = N_i E - TS , \qquad (2.35)$$

where S is the configurational entropy and is equal to klnW.
This leads to the following expression for μ $(=\partial F/\partial N_i)$

$$\mu_i = -kTln \left(\frac{1 - \alpha\theta_i}{\theta_i} \right) + E_i , \qquad (2.36)$$

which is reminiscent of eqn 2.5.

By decreasing the configurational degeneracy, site blocking has
the effect of giving a steeper μ_i/θ_i curve than for the ideal case
($\alpha = 1$). It should be recognized too that there is usually
mutual compensation of the entropy and enthalpy of solution such
that the blocking model often fits the activity quite well but
not the component partial entropy and enthalpy. A good example of this
is the fitting of blocking models to the carbon activity in
austenite (Lee 1974). Eqn 2.34 does not allow for the
simultaneous blocking of a given site by two or more defects.
Although Libowitz (1968) has commented that such refinements
are unnecessary at low defect concentrations, Monte Carlo
calculations by Oates et al. (1969) and Murch (1975b) have
indicated quite the contrary. There have been several analytical

refinements to eqn 2.34 to allow for multiple blocking (Moon 1963, McLellan et al. 1967 and Douglas 1964). Of these, only Douglas has proposed a procedure which avoids the inadequate simple combinatorial approach (McQuillan 1967). The treatment of Douglas is very specific to hydrogen interstitials in α-Zr and it is apparently difficult to extend the procedure to other cases (Oates et al. 1969, Gallagher et al. 1969).

A Monte Carlo procedure which exactly handles the problem of site blocking was first proposed by Baker (1966) in the analogous two dimensional problem of site blocking by adsorbed atoms on a surface. The procedure has subsequently been used by Oates et al. (1969) and Gallagher et al. (1969) to describe the configurational entropy associated with hydrogen dissolution in b.c.c. metals and carbon in austenite. Murch (1975b) has also used the method in his study of UO_{2+x}. Essentially, α is replaced by a petit canonical ensemble average $<\alpha>$ which is calculated from a computer lattice which has been brought to equilibrium (see section 3.5.3). Once the exact results are known, then one can test out various empirical analytical schemes (Boureau and Campserveux 1976, Hagemark 1963).

In the case of point defects, the number of nearest neighbour sites to be blocked is reasonably obvious. In the case of defect clusters such as the Willis (1964) defect cluster in UO_{2+x} and the Koch-Cohen (1969) cluster in $Fe_{1-\delta}O$, one can speculate fairly freely on the number of blocked defect cluster configurations and this has led in this case to α becoming something of an adjustable parameter (IAEA panel 1965, Roberts and Markin 1967, Libowitz 1968, Murch 1975b, Mitra and Allnatt 1979). Be this as it may, of the studies just cited, the first

three make use of a simple combinatorial equation of the form
of eqn 2.34. Murch (1975b) used a Monte Carlo method while
Mitra and Allnatt have used Mayer Cluster theory, an extension
to defect clusters of a previous treatment for point defects
(Allnatt and Cohen 1964). Mitra and Allnatt (1979) obtain good
agreement of their partial molar configurational entropy
with Monte Carlo results (Murch 1975b) in the case of 2:1:2
and 2:2:2 Willis defect clusters. We discuss Mitra and Allnatt's
work further in section 3.2.2 within the context of diffusion.

Inclusion of defect ionization is a natural extension to
ionic crystals of treatments dealing with a single defect type.
In UO_2, for example, hyperstoichiometry is associated with a
quite distinct valence state change of uranium from U^{4+} to U^{5+}.
We could write the oxidation reaction as

$$\tfrac{1}{2} O_2 (g) + 2 U_u^{\cdots} + V_i \rightleftharpoons O_i^{''} + 2U_u^{\cdots\cdots} . \qquad (2.37)$$

It is fairly clear that some of the oxidized cations remain in
the vicinity of the excess oxygen, possibly almost permanently
bound to it, while other oxidized cations are 'free'. In
eqn 2.37 we have written excess oxygen in UO_{2+x} as simple
interstitials but the same arguments apply to oxygen defect
clusters, fig. 2.3. One writes now for the petit canonical partition
function as exemplified by UO_{2+x}

$$Q = \frac{1}{N_i! \, N_5!} \sum_j \exp (-E_j/kT) , \qquad (2.38)$$

where N_5 is the number of oxidized cations.

In a Bragg-Williams treatment, one considers that the distribution of the oxidized cations is independent of the distribution of interstitial oxygen; both distributions are treated as being ideal. This leads to another term of the form: $-kT \ln \left(\frac{1-\theta}{\theta}\right)$ being added to the chemical potential of eqns 2.5 and 2.15. This has the effect, like site blocking, of generating steeper μ/θ_i isotherms. It can be seen, however, that some freedom can be permitted in the ratio of bound and unbound oxidized cations. In the case of defect clusters, this freedom, in combination with an adjustable number of blocked defect cluster configurations, as we noted above, probably permits the derived chemical potential to be fitted to almost any conceivable experimental isotherm! Treatments based on this are of somewhat doubtful value in the understanding of nonstoichiometry.

In a sophisticated treatment, the oxidized cations should be considered to be coulombically attracted to the primary or interstitial defect. In this case there should be no need to make a specific distinction between free and bound oxidized cations since there would then be a range of progressively bound oxidized cations, the distribution of which is temperature dependent. Although not quite proceeding that far, Atlas (1968a, b, 1970) made a very significant contribution to the statistical thermodynamics of nonstoichiometry by solving the partition function eqn 2.38 on the basis of coulombic interactions. Two levels of approximation were formulated. In the first, in applications to CeO_{2-x} and UO_{2+x}, Atlas assumed that the primary lattice defect (anion vacancy $- CeO_{2-x}$) and (2:1:2 interstitial cluster $- UO_{2+x}$ see fig. 2.3) interacted coulombically in a repulsive fashion. Similarly, the altervalent cations Ce^{3+} (CeO_{2-x}) and U^{5+} (UO_{2+x}) also interacted coulombically. Cross coulombic interaction (attraction) was ignored. Later, in another treatment of UO_{2+x}, Atlas (1970) included the contribution of cross interactions. In all cases the coulombic interactions were screened by means of a constant dielectric constant. Atlas wrote for the interaction potential ϕ

$$\phi = \frac{q_1 q_2}{r_{12}^2 \chi} \, , \qquad\qquad (2.39)$$

where q is the charge, r is the charge separation and χ is the dielectric constant. Although the dielectric constant is obviously variable on an atomic scale it may be represented by an average value. This value need not be the same as the static dielectric constant of the bulk because of the strong influence of interactions between close neighbours (Kröger 1964).

The basis of the Atlas treatment is (1) that a defect can control the occupancy of neighbouring sites within a small region of lattice and (2) that the defect potential energy is related to the local defect concentration. A lattice containing N_d defects is divided into small elements of equal volume each containing L lattice sites. The concentration of defects within each element depends on the interaction potential and the temperature. Defect energy levels denoted by u_0, u_1, u_2,... u_L are generated by reducing c, the reciprocal defect concentration, incrementally, giving reciprocal defect concentrations c, c-1, c-2,... c-L when each energy level contains n_0, n_1,... n_L defects. One now permits the defects to be placed *randomly* in any level, say, c-j, with the restriction that each defect removes c-j sites from the total number available. It can be seen that, in essence, the model of Atlas is, in fact, an ingenious site blocking scheme superimposed on a hierarchy of energy levels.

One can readily write down the following expression for the total number of ways, W, of randomly distributing N_d defects subject to the above conditions

$$W = \prod_{j=0}^{L} \frac{(c-j)^{n_j} w_j!}{n_j! (w_j - n_j)!} \, , \tag{2.40}$$

where $w_j = \dfrac{B + (c - j)n_j - \displaystyle\sum_{m=0}^{j} (c-m) n_m}{c-j}$. $\tag{2.41}$

The most probable values of $\{n_i\}$ can be found by maximizing $\ln W$ with respect to each n_i under the conditions

$$\sum_{i=0}^{L} n_i = N_d \tag{2.42}$$

and

$$\sum_{i=0}^{L} n_i u_i = <E'> \tag{2.43}$$

where $<E'>$ is the most probable value of the configurational energy due to the defects alone. This leads to the following equation for the chemical potential for the type of defect in question

$$\mu = kT \left[\frac{u_j}{kT} - \ln (c - j) + \ln \left(\frac{n_j}{w_j}\right) - \sum_{m=1}^{L} \frac{c - j}{c - m} \ln \left(\frac{1 - n_m}{w_m}\right) \right]. \tag{2.44}$$

Following the same procedure one can develop an expression analogous to eqn 2.40 for the counter defect. In the application of the first approximation to CeO_{2-x} and UO_{2+x} (Atlas 1968a, b) the agreement with experiment can only be described as fair. Nonetheless, the calculated compositional fluctuations of the partial molar entropy are indicative of a considerable persistence of residues of order in the nonstoichiometric phase as expected. With the second approximation in application to UO_{2+x} (Atlas 1970) the agreement was quite reasonable, Fig. 2.5. More recently, Murch (1973) solved the same model for UO_{2+x} with a Monte Carlo method (see section 3.5.3) and found somewhat better agreement with experiment, see Fig. 2.5.

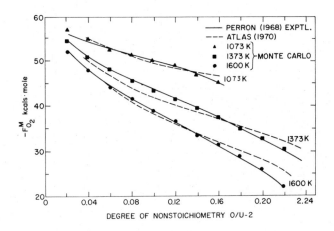

Fig. 2.5. The partial molar free energy of oxygen in UO_{2+x}. ————: Experimental data as accepted by Perron (1968), — — —: Atlas (1970), ▲ , ■ , ● : Monte Carlo (Murch 1973).

This study showed that the value of 5 assigned by Atlas to the dielectric constant is in fact low by at least a factor of 3. This low value of the dielectric constant probably compensates for the oversimplified configuration count in eqn 2.40.

Quite recently, Manes et al. (1976, 1979) used the statistics of Atlas to model CeO_{2-x} and PuO_{2-x}. They considered the formation of neutral tetrahedral complexes, each containing one oxygen vacancy and two reduced cations. This defect complex, or rather its dimer, has been deduced from structural studies as being a building block of the ordered phases found at low temperatures (Thornber and Bevan 1970). In the usual way, a characteristic formation energy can be associated with the complex. The interactions between the complexes were considered to be coulombic forces which arise from dipole moments induced in the complexes as a result of an internal transfer of charge. The configuration count of the complexes was handled as in eqn 2.40.

There comes, however, a limiting composition where the lattice becomes saturated with this type of complex. A larger complex must then be introduced in order to access higher defect concentrations. These complexes, with a higher formation energy, were also considered as neutral species. The interaction energies and the means to handle the distribution problem were treated in the same way as with the initial complex.

Ideally, of course, one would wish to handle this problem by treating both types of complex simultaneously and also include further species such as independent reduced cations and charged oxygen vacancies. This probably requires a Monte Carlo lattice gas approach along the lines initiated by Murch (1973). Ultimately, even this will be inadequate and it will probably be necessary to resort to Molecular Dynamics or fluid-type Monte Carlo calculations which, starting with realistic interatomic potentials, automatically should generate the correct defect species and distributions as a function of composition and temperature. Such calculations are not feasible on present computers except for very high temperatures.

Finally, it should also be recognized that differential thermodynamic functions such as the chemical potential, although a
sensitive probe of solid state structure and bonding, are not unique
in this respect, and any defect model must ultimately be compatible
with transport properties as well.

2.2.2.2 Interactions Between Defects (Cont.), Two Types of Defect

The treatments of the previous section have been suitable
for interstitial solid solutions or, in the case of a nonstoichiometric compound, large deviations from stoichiometry
where the presence of the complementary defect of, say, the Frenkel
pair is negligible. As we stated in section 2.2.1.2, to traverse
the homogeneity range between two diphasic regions one requires,
in general, two defect types, e.g., an anion vacancy and interstitial
Other combinations such as metal interstitial and anion interstitial
are also possible. There have been two formulations along these
lines: One due to Anderson (1946), the other due to Rees (1954).

The essential idea in the Anderson (1946) formalism is to
recognize an equilibrium between, say, anion vacancies and
interstitials. In effect, atoms are partitioned between regular
lattice sites and interstitial sites. The petit canonical partition
function is written as

$$Q = \frac{1}{N_i! \, N_v!} \sum_j \exp\left(-E_j/kT\right), \tag{2.45}$$

where the configuration energy of state j, E_j, is given in terms
of formation and nearest neighbour interaction energies

$$E_j = N_i E_i + N_v E_v + N_{ii}^j E_{ii} + N_{vv}^j E_{vv}, \tag{2.46}$$

where N_i and N_v are the number of interstitials and vacancies
respectively, E_i and E_v are the formation energies of an isolated
interstitial and vacancy, respectively, N_{ii} and N_{vv} are the
numbers of nearest neighbour interstitial-interstitial and
vacancy-vacancy pairs respectively in state j and E_{ii} and E_{vv}
are the corresponding nearest neighbour interaction energies.

Upon using the Bragg-Williams approximation, the maximum-term
method and the following definition of the two chemical potentials

$$\mu_i = \left(\frac{\partial F}{\partial N_i}\right)_{N_v, B, T} \quad \text{and} \quad \mu_v = \left(\frac{\partial F}{\partial N_v}\right)_{N_i, B, T} \, , \quad (2.47)$$

one finds that μ_i and μ_v are given by

$$\mu_i = -kT\ln\left(\frac{1-\Theta_i}{\Theta_i}\right) + [E_i + Z\Theta_i E_{ii}] \quad (2.48)$$

$$\mu_v = -kT\ln\left(\frac{1-\Theta_v}{\Theta_v}\right) + [E_v + Z\Theta_v E_{vv}]. \quad (2.49a)$$

The chemical potential of atoms on normal lattice sites, μ_n,
can be shown from eqn 2.49a, to be

$$\mu_n = -kT\ln\left(\frac{\Theta_v}{1-\Theta_v}\right) + [E_v + Z\Theta_v E_{vv}], \quad (2.49b)$$

where $\Theta_v = N_v/B$ and $\Theta_i = N_i/\gamma B$,

and B is the number of regular lattice sites and γB is the
number of interstices. At equilibrium, eqn 2.48 must equal eqn 2.49b.
The composite isotherm is found by graphical means. The result is
sketched schematically in fig 2.6 with E_{ii} and E_{vv} both < 0
(attraction). The theory thus relates the range of existence of
a nonstoichiometric compound to the interaction energies and

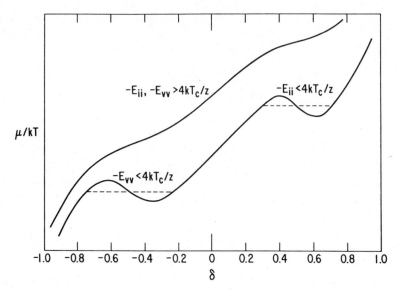

Fig. 2.6. A sketch of the composite isotherm from eqns 2.48 and 2.49b above and below the critical temperatures.

the defect formation energies. For $\theta_i << \theta_v$ or $\theta_v << \theta_i$, the chemical potentials are described by the appropriate *individual* equations 2.48 and 2.49b and we have accordingly the situation of a single defect type as described in section 2.2.2.1.

There have been rather few applications of the Anderson formalism. Thorn and Winslow (1966a) made use of it in $UO_{2\pm x}$. These workers also made use of the Stripp and Kirkwood (1954) formalism to calculate the vacancy vibrational contribution to the partition function. It was assumed in that formalism that the removal of N_v atoms from regular sites removed $3N_v$ frequencies of some characteristic value ν_v without otherwise altering the frequency distribution. With this model Thorn and Winslow found that the experimental activities in $UO_{2\pm x}$ are reasonably well reproduced. The formalism is, however, unable to reproduce the sigmoidal shape of the upper phase boundary. Thorn and

Winslow found, however, that the introduction of interstitial nearest neighbour blocking was sufficient to provide a fit to the phase boundary but this is inconsistent with the premise that $E_{ii} < 0$ for phase separation of U_4O_9 in the UO_{2+x} matrix! It was, in fact, Atlas (1970) who showed that phase separation of the *ordered* structure U_4O_9 can be generated by a model based on coulombic interactions between the interstitial defect clusters (fig 2.3) and oxidized cations (see section 2.2.2.1).

As we remarked above, the partition function, eqn 2.45 was solved at the Bragg-Williams approximation (Anderson 1946, Thorn and Winslow (1966a). There has been no attempt to solve this partition function at the level of the Quasi-chemical approximation although it would appear to this author to be relatively straight-forward to do so. The inclusion of longer range interactions and ionization of the defects similarly has not been essayed. A treatment which extends the Atlas (1970) formalism to both sides of stoichiometry is obviously called for.

Rather than start from the stoichiometric composition between two diphasic regions, Rees (1954) traversed the entire composition range by starting with the metal, A, and introducing interstitials, X, in such a way that as each interstial is introduced one or more interstices at higher energy are created nearby. In effect the problem now becomes one of partitioning atoms between two types of interstice subject to an energy difference between the sites (formally a difference in their formation energies) and interstitial-interstitial (of the same type) interaction energies. In order to generalize the idea, one considers that the number of sites of the i'th kind which are available for occupation depends on the number of occupied sites of the (i - 1)'th kind. For i = 2, Rees thus wrote down the following expression for the number of ways, W, of distributing N_i^1 interstitials of the first kind and N_i^2 of the second kind

$$W = \left(\frac{B!}{N_i^1! \ (B - N_i^1)!} \right) \left(\frac{N_i^1!}{N_i^2! \ (N_i^1 - N_i^2)!} \right) . \qquad (2.50)$$

Using the Bragg-Williams approximation for the calculation of the configurational energy the following parametric equations for the chemical potentials are obtained

$$\mu_i^1 = -kT\ln \left(\frac{1 - \theta_i^1}{\theta_i^1} \right) + E_i^1 + z\theta_i^1 E_{ii}^2 , \qquad (2.51)$$

$$\mu_i^2 = -kT\ln \left(\frac{\theta_i^1 - \theta_i^2}{\theta_i^1} \right) + E_i^2 + z(\theta_i^2/\theta_i^1) E_{ii}^2 , \qquad (2.52)$$

where $\theta_i^1 = N_i^1/B$ and $\theta_i^2 = N_i^2/B$ and the other terms retain the same significance as previously but are designated as being associated with the first or second type of site by the superscript 1 or 2 respectively.

At equilibrium eqn 2.51 must equal eqn 2.52 and again using graphical means we can obtain the composite isotherms. These have the same general form of fig 2.6 so we do not display them here. We should note, however, that the formalism has the capability of traversing any number of diphasic regions. The formalism has been applied with success to the Zr/H system (Martin and Rees 1954) and, more recently, by Murch and Thorn (1979e) to the Th/C system. Of particular interest in the formalism is the generation, as a result of eqn 2.52, of an asymmetric shape of the upper diphasic regions. This is, in fact, observed in the Th/C and U/C systems.

2.2.3. The Vibrational Partition Function

Thus far we have only considered the configurational aspects of the partition function. It is usual to add a vibrational component in the following way

$$Q = \frac{q^{N_i}}{N_i!} \sum_j \exp(-E_j/kT), \tag{2.53}$$

where q is the vibrational partition function of an interstitial. The above separability of the partition function into configurational and vibrational parts is a mathematical device which is reminiscent of the Born-Oppenheimer approximation. The two parts of the partition function should actually be correlated since the configurations of the defects are determined by the interaction energies; the difference in the potential energies of, say, two configurations must be offset by the differences between the kinetic energies associated with vibration (Thorn and Winslow 1967). One of the most outstanding problems today in the statistical thermodynamics of nonstoichiometry is the *correlation* of these configurational and vibrational partition functions.

In most modelistic work q is treated as either being equal to unity or as an unknown parameter which is absorbed eventually as a vibrational component of the partial molar entropy. Sometimes, however, q is expanded in the Einstein harmonic oscillator approximation to give

$$q = \frac{\exp(-\Theta/2T)}{1 - \exp(-\Theta/T)}, \tag{2.54}$$

where Θ, is the 'characteristic' temperature and $\Theta = h\nu/k$ where ν is the classical frequency of the order of 6×10^{12} hertz.

2.3. Nonstoichiometry approached from the ordered state –

Microdomain Theories

An important aspect of recent structural work in nonstoichiometry is the existence of defect clusters or microdomains such as the Willis (1964) clusters in UO_{2+x} (fig 2.3) and the Koch-Cohen (1969) cluster in $Fe_{1-\delta}O$ (fig 2.4). Even before these studies Ariya and coworkers (1958, 1962) had proposed a theory of nonstoichiometry based on thermochemical arguments, which described nonstoichiometric compounds as consisting of dispersed microdomains of an adjacent phase. To support his conjecture, Ariya noted that the enthalpy of formation of, say, MO_{1+x} can be expressed as a linear combination of the formation enthalpies of MO and M_2O_3

$$\Delta H_f^\circ \ (MO_{1+x}) = (1 - 2x)\Delta H_f^\circ MO + 2x\Delta H_f^\circ \ (M_2O_3). \tag{2.55}$$

While this implies a constant heat of solution in the non-stoichiometric phase which is not, in fact, usually observed in this more structure-sensitive quantity, the concept is nonetheless significant. The contiguous intergrowth of microdomains in a matrix probably is the result of two principles operating at the atomic level (Thorn 1970). The first is the local neutralization of charge. The second is a minimization of the appropriate thermo-dynamic energy function at the mismatched interface between the microdomain and the matrix. Both of these principles act to stabilize the degree of nonstoichiometry in the most energetically favourable manner.

Ubbelohde (1957, 1966) has proposed that the strain and surface energy of a microdomain could add further degrees of freedom to the system. Thus, a single phase system containing microdomains

may just as well be considered a pseudo-two-phase mixture which
exhibits bivariant behaviour. As has been pointed out on several
occasions (O'Keeffe 1970 and Murch 1975), interacting 'defects'
or defect aggregates may then be considered fragments or structural
residues of the neighbouring phase rather than a solute in a solid
solution.

Anderson (1969, 1970) has treated the dispersion of microdomains in terms of ideal solution theory. The basis of the
treatment was a relation of the *size* of the microdomain to an
excess chemical potential. Anderson inferred from comparison
of his relations with the experimental activity isotherms in
CeO_{2-x} (Bevan and Kordis 1964) that any ordered element is
probably biassed towards the smallest unit.

Thorn (1970) has sketched what seems to us to be sufficient conditions
for a microdomain theory. Thorn took an approach reminiscent
of significant structure theory of liquids (Eyring and Marchi 1963).
Defining a correlation function $\rho(\mathcal{N})$ which is an explicit function
of the *total* number of defects, Thorn wrote down a *joint*
probability density function of the form

$$f = F^{\rho} . G^{1-\rho} \qquad\qquad (2.56)$$

where F and G are the individual probability density functions of
the microdomain and matrix respectively.

To the extent that ρ is continuous, f represents a pleasing situation
wherein the matrix is continuously converted into microdomain
and vice versa. Assuming random mixing of microdomains and
other conditions which seem by the author's admission rather

obscure one can generate from this formalism the component isotherms as found by Anderson (1946) (eqns 2.48 and 2.49b). To this extent, the approaches from the disordered and ordered states have converged. In essence, what has been demonstrated so far is a consistency, at the random mixing level, of the two approaches. The conceptual approach by Thorn is, however, a significant step forward in a generalized theory of nonstoichiometry and one which, in this author's opinion, is likely to prove an incentive to others.

3. Atomic Transport

3.1 Introductory Remarks

A description of transport in highly defective solids has
had a relatively short history when compared to the
corresponding description of equilibrium properties. Despite
this, considerable progress along rigorous lines has now been
made, and, with the continued interest in these materials
for technological applications, such progress is likely to
continue.

Much of the early work, however, was incorrectly
formulated, mainly as a result of attempts to extend the
random walk formalism, as developed for metals and stoichiometric
binaries, into the regime of solids with high defect concentrations.
It was rarely appreciated, for example, that the Bardeen and
Herring (1952) tracer correlation factor could depend on
defect concentration, or that a high concentration of defects
led to a situation wherein the atomic jump frequency could
vary from site to site rather like a partially ordered alloy.
Other difficulties arose from arbitrary and often inconsistent
definitions of the defect concentration.

We have chosen, therefore, not to discuss the early history
of the field. Rather, we have aimed to present later theoretical
developments, but in a manner which emphasizes their consistency,

their evolution via models of increased complexity and, of course, their agreement with experimental observations. We cover three important areas of diffusion: tracer diffusion, ionic conductivity and chemical diffusion, primarily from a statistical mechanical viewpoint. Later we discuss several recent theoretical methods which have been primarily responsible for our current level of understanding of atomic transport in highly defective solids.

For reference, we have defined the various diffusion co-efficients used in the text in appendix I. In appendix II we discuss some of the experimental techniques in tracer diffusion that have been developed in *recent* years and which have had, or are likely to have, a substantial impact on the acquisition of experimental data.

3.2 Tracer Diffusion

3.2.1 Non-interacting Defects

3.2.1.1 One Type of Defect

Tracer diffusion in a solution of point defects or ions of a *single* type which do not interact is probably the simplest case to consider. As we noted in section 2.2., the case of a single type of defect is appropriate for diffusion in interstitial solid solutions, in the metallic nonstoichiometric carbides, nitrides etc. and in the intercalation compounds like Li_xTiS_2. It is implied that in the nonstoichiometric compounds referred to above the composition is sufficiently far from stoichiometry that the influence on diffusion of the complementary defect of, say, the Frenkel pair, is negligible.

We start with the following very general equation for the tracer diffusion coefficient, D* in a system with a single mechanism of migration

$$D* = \tfrac{1}{2} \Gamma\lambda^2 f, \tag{3.1}$$

where λ is the component of the jump distance in the diffusion direction, Γ is the atomic jump frequency and f is the tracer correlation factor.[†] Both Γ and f are strongly dependent on defect concentration; for a single mechanism, λ depends on concentration only through the expansion or contraction of the lattice.

The Jump Frequency

The decomposition of Γ to its components such as the defect concentration requires a little care. We recommend the following procedure to ensure self consistency. The total number of defect jumps in some time, t, is given by $N_d Z\omega t$ where N_d is the number of defects, Z is the number of saddle points available to the defect and ω is the atom-defect exchange rate. On the other hand, the number of atom jumps in time t is given by $N_a \Gamma t$, where N_a is the number of atoms. Within the same time, t, the total number of atom jumps must equal the total number of vacancy jumps. Accordingly, Γ becomes

$$\Gamma = N_d Z\omega/N_a. \tag{3.2}$$

[†]For an excellent and extensive review on correlation effects in solids, see LeClaire (1970).

The relation 3.2 is, of course, valid under any circumstances and not just under the condition of no interactions. For the noninteracting case ω may be expanded to

$$\omega = \nu\exp\ (-G^{\circ}_{act}/kT)\,\phi\ ,\qquad\qquad (3.3)$$

where ν is the fundamental lattice vibrational frequency, G°_{act} is the free energy of migration of an isolated atom and ϕ is the probability of finding an atom neighbouring to a defect.

If we define a vacancy concentration; $\Theta_v = N_v/B$ where N_v is the number of vacancies and B is the number of regular sites, then ϕ is given by

$$\phi = (1-\Theta_v),\qquad\qquad (3.4)$$

and Γ reduces to

$$\Gamma = \Theta_v Z\nu\exp\ (-G^{\circ}_{act}/kT).\qquad\qquad (3.5a)$$

It is often convenient to expand G°_{act} as

$$G^{\circ}_{act} = H^{\circ}_{act} - TS^{\circ}_{act},\qquad\qquad (3.5b)$$

where H°_{act} and S°_{act} are the activation enthalpy and entropy respectively. H°_{act} is usually approximated by E°_{act} in statistical mechanical (constant volume) treatments.

In the case of the interstitialcy mechanism, with no interactions Γ is given by

$$\Gamma = \frac{\Theta_i}{\gamma^{-1}+\Theta_i}\ Z\nu\exp\ (-G^{\circ}_{act}/kT),\qquad\qquad (3.6)$$

where $\Theta_i = N_i/\gamma B$ and N_i is the number of interstitials and γ is the ratio of interstices to regular sites.

Tracer Correlation

The concept of tracer correlation in diffusion was introduced
by Bardeen and Herring, surprisingly as late as 1952. The origin
of the tracer correlation effect lies in the non-random motion
of a tracer atom which is induced by a defect. For an excellent and
extensive review of the correlation factor and the various
'random walk' methods of calculation we refer the reader to
LeClaire (1970). In most of the older literature on correlation
it is assumed, either implicitly or explicitly, that the defect
concentration is very low. The variation of the tracer correlation
factor with defect concentration and defect - defect interactions
has, in fact, only been discussed within the last decade. To
appreciate the influence of the first of these effects on f
we refer to figure 3.1. Here, we consider the motion of a tracer
atom (shaded) at position B relative to a vacancy at A. Let us
assume that the tracer and the vacancy have just exchanged places.
Because the vacancy is still adjacent to the tracer atom, the
tracer may subsequently reverse its last jump at a probability
higher than for a jump elsewhere. If a new vacancy[†] appears at
C, the probability of that reverse jump decreases. In effect, the
new vacancy has decorrelated the motion of the tracer atom.
Consequently, as the vacancy concentration θ_v rises the tracer
correlation factor rises from its geometric value, f_0, at an
infinitely low vacancy concentration to the value of unity as
$\theta_v \rightarrow 1$. This terminal value signifies an uncorrelated random walk
at the limit of a single atom remaining. This is, of course the
value of the tracer correlation factor for interstitial diffusion
at an infinitely dilute solution of interstitials.

[†]For the purposes of the illustration the vacancies are deemed
distinguishable.

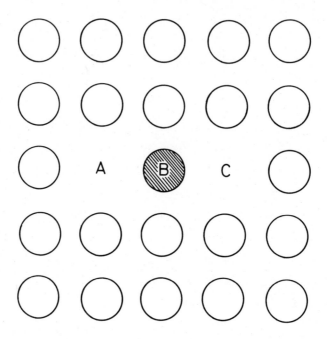

<u>Fig. 3.1.</u>   Illustration of the decorrelation of a reverse jump B → A
of a tracer by the presence of a vacancy at C.

The functional form of the dependence of the tracer
correlation factor on vacancy concentration in the well-known
lattices has never been calculated rigorously by an analytical
means. Recently, there have, however, been a number of approximate
treatments which we will now briefly disucss.

In the course of their important work concerning the application
of the Path Probability (PP) method to highly defective solids,
see section 3.5.2,Sato and Kikuchi (1971 a, b) deduced the
following linear expression for f in the honeycomb lattice

$$f(\Theta_v) = f_o(1-\Theta_v) + \Theta_v. \qquad\qquad (3.7)$$

More recently, Benoist et al., (1977) have presented a new method of calculating correlation factors. The method, which makes use of double LaPlace - Fourier transforms, provides an approximate solution to several outstanding correlation problems, and in particular the dependence of f on vacancy concentration.

Fedders and Sankey (1978), using the multiple scattering approximation, arrived at the following expression for f in the simple cubic lattice

$$f(\Theta_v) = (1-2\alpha)/[1+(2-6\Theta_v)\alpha/(1-\Theta_v)] , \qquad (3.8)$$

with $\alpha \simeq 0.1059$.

Finally, Haven (1978) deduced the following approximate form for f

$$f(\Theta_v) = \frac{1 + \beta<\cos\Phi>}{1 - \beta<\cos\Phi>} \qquad\qquad (3.9)$$

with $\beta = [1-\Theta_v]/[1+(Z-1)\Theta_v]$

where $<\cos\Phi>$ is the average value of the cosine of the angle between any two consective jump vectors at vanishingly small Θ_v.

There have, however, been quite a few Monte Carlo (MC) studies of the functional dependence of f on Θ_v. For details on the MC method we refer to section 3.5.3. DeBruin and Murch

(1973) and Murch (1973) calculated f in the simple cubic lattice.
Both eqns. 3.7 and 3.8 have been said to agree with the simulated
values to within about 1% over the entire composition range.
The results of Benoist et al. (1977) also agree within about
1% up to a concentration Θ_v = 0.2. Higher vacancy concentrations
were considered to be outside the limits of their method. No
detailed comparison has been made of the MC results with Haven's
equation – (3.9).

The MC method has also been applied to the honeycomb
(Murch and Thorn 1977a) and the f.c.c. lattices (Murch 1975a).
In each case eqn 3.7 is a quite reasonable approximation and,
to the extent of an application to a real system of an approximate
model based on zero defect-defect interactions, the author
recommends this equation.

In the case of the *interstitialcy* mechanism, however, much
less work has been done on f as a function of interstitial
concentration. DeBruin and Murch (1973) and Murch (1973) performed
a single MC calculation of the correlation factor for interstitialcy
diffusion, f_i^{cy}, in the square planar lattice. In this case the
correlation factor remains essentially constant for up to about 30%
interstitials, then decreases slowly to the value for f_o for vacancy
diffusion in the new lattice derived from the regular packing of
interstitials in the original lattice. In this instance, of course,
two atoms move cooperatively toward a vacancy, but to first order
this does not change the value of f_o from that for the *single*
atom/vacancy exchange mechanism.

Recently, Haven (1978) deduced the following linear equation for $f_i^{cy}(\theta_i)$

$$f_i^{cy}(\theta_i) = (1-\theta_i)(1 + < \cos \Phi >) \qquad (3.10)$$

This equation would seem to be in only semi-quantitative agreement with the results of the computer simulation described above.

There have been several *applications* to real systems which have involved calculations of f in the noninteracting defect approximation. Murch (1975a) investigated carbon tracer diffusion in UC_x, 1.0<x<2.0. In the rock-salt like structure of UC_x excess carbon is accomodated by double occupancy of the octahedral interstices. In the generally accepted mechanism of carbon migration (Murch and Thorn 1977b, Catlow 1976) it is presumed that the acetylide - like di-interstitial splits, with one of the released atoms migrating to a nearest neighbour singly occupied site while the other released atom is presumed to relax in order to assume residency as a single occupant. Such a mechanism can reasonably be approximated by the vacancy mechanism provided that the C_2 group is not a free rotator. In the course of an investigation of carbon diffusion in γ Fe (austenite) Murch (1979) used eqn 3.7 as an approximation for the functional dependence of f in the f.c.c. lattice. Bustard (1979) investigated hydrogen diffusion in $\gamma-TiH_x$, 1.5<x<2.0. It was assumed in that study that migration of hydrogen took place by third-nearest-neighbour hopping in a simple cubic lattice. This corresponds to nearest neighbour hopping in the diamond lattice. Equation 3.7 was again used to approximate $f(\Theta_v)$ in that system. Finally, Haven (1978) used eqn 3.10 with $\Theta_i = 0.158$ in a theoretical estimate of the Haven ratio (see section 3.3.2) in β-alumina.

3.2.1.2.   Two Types of Defect

In nonstoichiometric compounds close to the stoichiometric composition it is often necessary to include the contribution to atomic transport of the complementary defect of the defects pair. In the spirit of the assumption of non-interacting defects one may permit each defect *type* to contribute independently to diffusion. That is to say, defects within each type do not interact nor do the two types of defect cross-interact. Under these conditions it is permissible to write the tracer diffusion coefficient as the sum of two parts

$$D^* = \tfrac{1}{2}\, \Gamma_v \lambda_v^2 f_v \;\; + \;\; \tfrac{1}{2}\, \Gamma_i \lambda_i^2 f_i^{cy}, \tag{3.11}$$

where the subscript v symbolizes diffusion via vacancies and the subscript i symbolizes diffusion via interstitials and f_i^{cy} signifies the tracer correlation factor for the interstitialcy mechanism

Molecular Dynamics studies in ionic solids such as $SrCl_2$ and CaF_2 (Rahman 1976 and Gillan and Dixon 1980) and lattice gas calculations in β-alumina (Sato and Kikuchi 1971 and Murch and Thorn 1977e) strongly suggest that a clear distinction cannot always be made between intrinsic diffusion mechanisms and the process of creation and annihilation of Frenkel pairs. When there is such coupling, D^* cannot be written rigorously in the form of eqn 3.11.

According to eqns 3.5 and 3.6 the atomic jump frequencies contain the defect concentrations which are related through the Frenkel product

$$\frac{\theta_i \theta_v}{(1-\theta_i)(1-\theta_v)} = \exp(-G_F/kT),$$ (3.12)

and G_F is the free energy of formation of an isolated Frenkel pair. At the stoichiometric composition in a compound MO

$$\theta_i \gamma = \theta_v.$$ (3.13)

Eqn 3.13 may then be substituted into the expressions for Γ_v and Γ_i (eqns 3.5 and 3.6) and these are finally substituted into eqn 3.11 and 3.12. This is a familiar procedure well-known from ionic conductivity measurements on, say, alkaline earth fluorides (Barsis and Taylor 1966).

However, for nonstoichiometric compounds, and we now use UO_{2+x} as a concrete example, we wish to know the dependence of D^* for oxide ions on *composition*, x (Murch and Thorn 1978a). In our example, $\gamma = \frac{1}{2}$ and composition, x, θ_i and θ_v are related by

$$x = \theta_i - 2\theta_v.$$ (3.14)

From eqns 3.12 and 3.14 we can now deduce the following expressions for θ_i and θ_v in terms of x (Contamin et al., 1972, Matzke 1966 and Lidiard 1966)

$$\theta_i \simeq \frac{1}{2}[x + \sqrt{x^2 + 8\exp(-G_F/kT)}]$$ (3.15)

and

$$\theta_v \simeq -\frac{1}{4}[x - \sqrt{x^2 + 8\exp(-G_F/kT)}]$$ (3.16)

Eqns 3.15 and 3.16 may now be substituted into the expressions
for Γ_i and Γ_v (eqns 3.5 and 3.6) and hence eqn 3.11 for D^* is
finally expressed in terms of x. For completeness eqns 3.15 and
3.16 should also be used in the expressions for the correlation
factor, eqns 3.7 and 3.10.

Interestingly, we can obtain exactly the same expressions for
θ_i and θ_v (eqns 3.15 and 3.16) by solving the configurational
partition function for the nonstoichiometric crystal at the
Bragg-Williams approximation (see section 2.2.2.2). This was the
procedure taken by Thorn and Winslow (1966a,b). At close examination,
such equivalence is not surprising since the assumption of ideal
mixing of defects which is implicit in the form of the Frenkel
product (eqn 3.12) is also the principal assumption of the Bragg-
Williams approach. As we discussed in section 2.33 Thorn and
Winslow (1966a) also included in their total partition function
for UO_{2+x} defect contributions to the separable vibrational part.
The contribution of interstitials was termed q_i and of the vacancies
q_v. If we carry these contributions through the analysis we find
the following expressions for θ_i and θ_v analogous to eqns 3.15
and 3.16

$$\theta_i \simeq \tfrac{1}{2} \left[x + \sqrt{x^2 + \frac{8q_i}{q_v} \exp\,(-E_F/kT)} \right] \qquad (3.17)$$

and

$$\theta_v \simeq -\tfrac{1}{4} \left[x - \sqrt{x^2 + \frac{8q_i}{q_v} \exp\,(-E_F/kT)} \right] \qquad (3.18)$$

where E_F is the *energy* of the Frenkel pair formation. The formation entropy of the Frenkel pair has been absorbed into q_i/q_v. In $UO_{2\pm x}$ the ratio q_i/q_v is of the order of 60 and varies slowly with temperature.

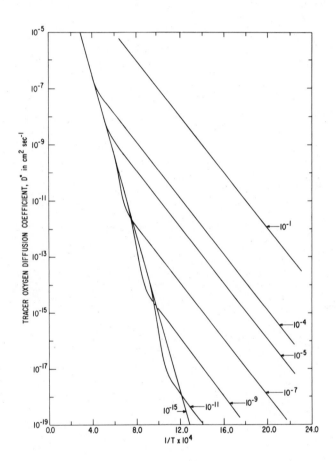

Fig. 3.2. The calculated temperature dependence of D* at various values of composition x in UO_{2+x}.

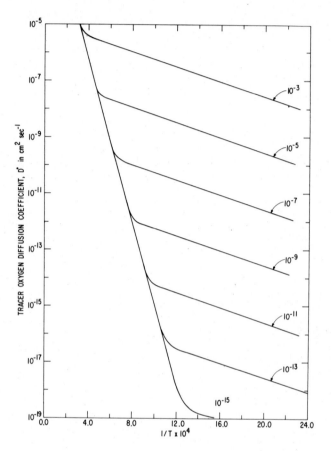

Fig. 3.3. The calculated temperature dependence of D* at various values of composition x in UO_{2-x}.

The results for $UO_{2\pm x}$ after using eqns 3.17 and 3.18 in the equation for D^* (eqn 3.11) are displayed in figs 3.2 and 3.3 (Murch and Thorn 1978a). Considering UO_{2+x} first, we see that at low temperatures and high compositions the activation energy for diffusion represents only the enthalpy of migration of *interstitials*. Similarly, for UO_{2-x}, for high compositions and low temperatures, the activation energy represents only the migration enthalpy of *vacancies.* In both cases, at lower compositions and higher temperatures these families of curves converge onto a single 'stoichiometric' curve wherein the activation energy is the sum of half of the Frenkel energy and the lower migration enthalpy of the two defects. It also turns out that on one side of stoichiometry (in our case UO_{2+x}), the family of curves inevitably converges onto the high temperature curve through a transition region wherein the slope attains a value *greater* than either of the high or low temperature branches. In the case of UO_2 the migration energy for vacancy diffusion E_v^M is less than E_i^M (Catlow et al., 1975), the migration energy for interstitialcy diffusion. The slope of the high temperature curve is

$$E_{act} = E_v^M + \tfrac{1}{2}E_F. \qquad (3.19)$$

On the hypostoichiometric side the family of curves at low temperature give

$$E_{act} = E_v^M. \qquad (3.20)$$

On the hyperstoichiometric side the family of curves at low temperature give

$$E_{act} = E_i^M. \qquad (3.21)$$

The transition region is on the hyperstoichiometric side of our example and E_{act} can attain a value given by

$$E_{act} \leqslant E_v^M + E_i^M + \tfrac{1}{2}E_F. \qquad (3.22)$$

We know of no system[†] where the transition region has in fact been detected. The width of the transition region shrinks at high

---

[†]Conductivity studies in silver halides reveal similar transition regions (Corish and Jacobs 1972) but the concentration of defects in this and the 'extrinsic' region is controlled by impurities.

temperatures, in fact, at the temperature of diffusion measurements it may well be almost impossible to detect. Moreover, the control and analysis of a constant composition very close to the stoichiometri composition is a most difficult task. In practice, any composition, x, $\lesssim 10^{-6}$ will be indistinguishable from the stoichiometric composition and inspection of Fig. 3.2 indicates that the transition region even at a composition of 10^{-6} could easily be overlooked. Other problems concern the interpretation of the as-measured composition when immobile cation vacancies are present from a previous high temperature heat treatment (Thorn and Winslow 1966c) or when a high level of impurities is present (Kofstad 1972).

3.2.1.3 Mass Action Law Approach

Principally because of the difficulty involved in maintaining a composition constant in a series of diffusion experiments at different temperatures, it has become common to perform experiments at constant pressure of the more volatile component (see the treatise by Kofstad (1972)). In particular, many nonstoichiometric oxides have been investigated in this manner. Some physics is concealed this way because (1) there is an essential cancellation of defect interaction terms, see below, and (2) the results can really only be interpreted in the light of the mass action laws approach.[†] This approach, as usually applied, is based on non-interacting defects, i.e., a constant heat of solution and an ideal entropy of mixing. Some attempt can be made to introduce defect interactions by introducing the association of defects but this seems to us an *ad hoc* approach (Kröger 1964).

[†] If composition/pressure isotherms are available, then the functional dependence of D* on composition can be determined. Otherwise, the link can be made only through a model.

Charged defects can readily be introduced into the formalism of mass action laws but only within the framework of ideal mixing, surely a poor assumption under the circumstances. The atomic jump frequency Γ is still judged to have the same form as eqns 3.5 and 3.6. That is to say, correlation between the motion of the various charged species is considered nonexistent.

We exemplify discussion of the mass action laws approach by considering an oxide MO_{1+x} capable of being defective on either side of stoichiometry. We write for the reaction of the removal of a normal atom from the solid to the gas phase

$$O_O \; \rightleftharpoons \; V_O^x \; + \; \tfrac{1}{2}O_2.$$ (3.23)

This leads to an expression for the equilibrium constant, $K_{V_O^x}$

$$K_{V_O^x} = \frac{\left[V_O^x\right] p_{O_2}^{\frac{1}{2}}}{\left[O_O\right]},$$ (3.24)

and for small deviations from stoichiometry

$$\left[V_O^x\right] \alpha \; p_{O_2}^{-\frac{1}{2}}.$$

Since $\left[V_O^x\right] = \theta_v$, then Γ of eqn 3.5 and hence D^* will be proportional to $p_{O_2}^{-\frac{1}{2}}$. We may also consider ionization reactions of the form

$$V_O^x \; \rightleftharpoons \; V_O^{\cdot} + e'.$$ (3.25)

$$V_O^{\cdot} \; \rightleftharpoons \; V_O^{\cdot\cdot} + e''.$$ (3.26)

In nonstoichiometric ionic compounds, the electrons are considered to be localized at cations i.e., 'valence' defects and eqns 3.25 and 3.26 are written as

$$M_M + V_O^X \rightleftharpoons V_O^{\cdot} + M_M', \qquad (3.27)$$

and
$$M_M + V_O^{\cdot} \rightleftharpoons V_O^{\cdot\cdot} + M_M'. \qquad (3.28)$$

Eqns 3.27 and 3.28, when combined with the corresponding equations of electroneutrality

$$\left[M_M'\right] = \left[V_O^{\cdot}\right] \qquad (3.29)$$

and
$$\left[M_M'\right] = 2\left[V_O^{\cdot\cdot}\right], \qquad (3.30)$$

lead to pressure dependencies of $p_{O_2}^{-\frac{1}{4}}$ and $p_{O_2}^{-1/6}$ respectively for D^*.

More generally, one wishes to traverse from one side of the stoichiometric composition to the other. We may introduce the reaction

$$V_i + \tfrac{1}{2}O_2 \rightleftharpoons O_i. \qquad (3.31)$$

O_i may ionize thus, assuming localization of the electron hole at a cation

$$O_i + M_M \rightleftharpoons M_M^{\cdot} + O_i' \qquad (3.32)$$

and
$$O_i' + M_M \rightleftharpoons O_i'' + M_M^{\cdot}. \qquad (3.33)$$

This leads to pressure dependencies of $p_{O_2}^{\frac{1}{4}}$ and $p_{O_2}^{1/6}$ respectively for D^* (via eqn 3.6 assuming $(\gamma^{-1} + \Theta_i) \simeq 1$). Focussing attention on, for example, doubly charged defects, we may invoke the Frenkel product, which is written for small deviations from stoichiometry, c.f. eqn 3.12

$$\left[O_i'' \right] \left[V_O^{\cdot\cdot} \right] \; = \; \exp \, (-G_F/kT) \; = \; K_F, \tag{3.34}$$

and intrinsic ionization:

$$M_M \; \underset{\longleftarrow}{\overset{\longrightarrow}{\rule{1cm}{0pt}}} \; M_M' + M_M^{\cdot} \tag{3.35}$$

with $\qquad K_I = \left[M_M' \right] \left[M_M^{\cdot} \right].$

The behaviour of $\left[V_O^{\cdot\cdot} \right]$ and $\left[O_i'' \right]$ at the stoichiometric composition is of particular interest. If intrinsic ionization predominates:

$$\left[M_M' \right] \; = \; \left[M_M^{\cdot} \right] \; = \; K_I^{\frac{1}{2}} \; >> \; \left[V_O^{\cdot\cdot} \right] \; \text{and} \; \left[O_i'' \right] \tag{3.36}$$

and M_M' and $M_M^{\cdot}$ will be independent of p_{O_2} while $\left[O_i'' \right]$ is proportional to $p_{O_2}^{\frac{1}{2}}$ and $\left[V_O^{\cdot\cdot} \right]$ is proportional to $p_{O_2}^{-\frac{1}{2}}$.

Conversely, if

$$\left[O_i'' \right] \; = \; \left[V_O^{\cdot\cdot} \right] \; = \; K_F^{\frac{1}{2}} \; >> \; \left[M_M' \right] \; \text{and} \; \left[M_M^{\cdot} \right] \tag{3.37}$$

then $\left[O_i'' \right]$ and $\left[V_O^{\cdot\cdot} \right]$ are independent of p_{O_2} while the valence defects $\left[M_M^{\cdot} \right]$ and $\left[M_M' \right]$ are proportional to $p_{O_2}^{\frac{1}{4}}$ and $p_{O_2}^{-\frac{1}{4}}$ respectively.

Nonstoichiometric ionic solids are characterized by the ease by which electronic intrinsic disorder takes place. The case above where intrinsic electronic equilibrium predominates is therefore the one normally encountered. In fig. 3.4 we schematically represent the concentrations of O_i'', $V_O^{\cdot\cdot}$, M_M' and $M_M^{\cdot}$ as a function of partial pressure of oxygen. It would now be possible to determine, with the aid of eqn 3.5, 3.6, and 3.11, the pressure dependence of D^* as the entire range of homogeneity is traversed. To the author's knowledge this has never been attempted.

58

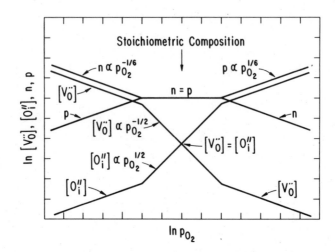

Fig. 3.4. The concentration of point defects and electron carriers for an MO oxide containing Frenkel disorder in the oxygen ion sublattice versus oxygen pressure (after Smeltzer and Young (1976)).

The foregoing equations are traditionally used in the following way. Experimentally, the slope of a $\ln D^*/\ln p_{O_2}$ plot is determined and this is said to specify the defect type and charge state. From a statistical mechanical viewpoint, this is quite equivalent to fitting a particular *point* defect model, with the following assumptions, to the p_{O_2}/composition plot. One neglects the variation of the heat of solution and assumes ideality for the entropy of solution. The adjustable parameters in the fitting of this plot are then the number of defect types and their relative concentrations subject to the condition of electroneutrality.

We now point out what may seem an inconsistency in our discussion. The dependence of D^* on x through eqns 3.15 and 3.16 substituted into eqn 3.11 was also the result of ideal mixing yet evolved from a partition function eqns 2.1, 2.13 containing defect-defect interaction terms. The inconsistency comes about because of the nature of the approximation for the solution of that partition function. At the BW approximation the defects are treated as interacting but nonetheless ideally mixed. The interaction terms are then *not* transmitted to $\Theta_i(x)$ and $\Theta_v(x)$, but rather, are transmitted only to $\Theta_i(p_{O_2})$ and $\Theta_v(p_{O_2})$. Alternatively, we may think of the interaction terms as being transmitted to Γ_i and Γ_v (eqns 3.5 and 3.6) in a way which we shall discuss in section 3.2.3. Accordingly, $D^*(p_{O_2})$ at the Bragg-Williams approximation would contain defect interaction terms in contradistinction with the result of a mass-action-law approach. In actual fact these interaction terms essentially cancel out and we are left with the same result as the mass-action-law approach with the physics of non-ideality effectively concealed, we refer to section 3.2.3 and Murch (1980d).

3.2.2 Tracer Diffusion Involving Defect Clusters

As a bridge between the theories of noninteracting (section 3.2.1) and interacting (section 3.2.3) defects we can now introduce the concept of diffusion in a random solution of defect clusters. The formation and structure of defect clusters in nonstoichiometric phases is discussed in some detail in sections 2.3 and 2.2.2.1.

Firstly, we assume a migration mechanism which does not break up the cluster. While the cluster may change its orientation and internal configuration it still is assumed to retain a characteristic identity after each cluster jump. The longevity of a given cluster may be shortened, however, if the clusters interact in such a way that cluster chains of variable length are annihilated or created. Now, without direct experimental evidence it is somewhat idle to speculate too freely on the cluster migration

60

mechanism. Nonetheless, valuable insight into tracer correlation can be gained by investigating possible mechanisms. Greenwood and Howe (1972) have speculated on a migration mechanism of the Koch-Cohen (1969) cluster in $Fe_{1-\delta}O$. DeBruin and Murch (1973) and Murch (1973) have speculated similarly on a migration mechanism of the Willis (1964) 2:2:2 cluster in UO_{2+x}. (This mechanism is illustrated in fig 3.5). In the latter study, the authors were able, in a Monte Carlo investigation, to calculate a tracer correlation factor for the simultaneous diffusion of four inequivalent nonsuperimposable clusters. The calculation revealed a result which was at once surprising and yet fascinating; the correlation factor took a value of 1.587 which decreased slowly with concentration of defect clusters. A repeat calculation by Klauber (1977) with an entirely new program has verified the result. This seems to be the only case in the solid state diffusion literature where a calculated correlation factor has taken a value greater than unity.

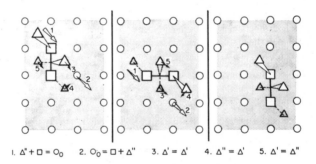

Fig. 3.5. Postulated jump sequence for the coherent diffusion of a Willis (1964) defect cluster. The plane of the drawing is (100), (after Murch (1973) and de Bruin and Murch (1973)).

It has not been possible as yet to determine precisely why such an interesting result emerges. Despite the enormous power of the Monte Carlo method, the technique becomes somewhat disappointing and extremely tedious when the finer features of the jump sequence need to be probed. For this reason, a somewhat simpler mechanism which gives a similar result should be sought.

However, there seems no better place than here for the author to express briefly his *opinion* on the f>1 question. The cluster mechanism of fig 3.5 may be envisaged as a mechanism which operates on tracer atoms rather like a driving force which sweeps the atoms along for several jumps. This can occur because a *given* defect cluster itself migrates in quite an anisotropic manner. With respect to a tracer atom the driving force changes direction after several tracer jumps when the tracer atom may leave[†] the first cluster and become part of another cluster of perhaps a different configuration (there are four inequivalent configurations). As a given tracer atom encounters the various different clusters, it undergoes a series of displacements which are quite anisotropic in each case. But over many atom/cluster encounters the tracer atom appears to undergo *isotropic* diffusion. Exploratory computer studies by the author on simple biassed random walks, though not exactly analogous to the cluster mechanism above, certainly indicate that an *apparent* tracer correlation factor, greater than unity, can in fact be built up this way. In brief summary then, in the author's opinion, a greater than unity value for the tracer correlation factor requires a unique set of defect clusters, *each* of which moves anisotropically, but which produce isotropic diffusion with respect to a tracer atom which interacts with the *complete* set. This explanation is

---

This step, and those preceeding it, must each be correlated in a way like the interstitialcy mechanism (LeClaire 1970).

faintly analogous to that describing the structure of the crystal
containing defect clusters: despite the local assymmetry of a
given cluster, in a *randomized* configuration the average translational symmetry *remains* that of the host crystal. In the case
of UO_{2+x} the symmetry of the phase remains cubic despite the
non-cubic symmetry of the individual clusters.

At the level of no interactions between defect clusters we
now discuss the tracer diffusivity assuming diffusion via clusters.
Eqn 3.2 is still relevant to such a situation but the calculation of
the probability , ϕ, that an atom can be found neighbouring to a defect
cluster in the sense that it is possible for the atom to 'enter'
the defect cluster in the next cluster jump, is, however, a
non-trivial problem. The reason for this is that it is not
possible to have ideal mixing of defect clusters in the same sense
as it is with point defects. That is to say, because each defect
cluster occupies several lattice sites each cluster in effect
removes more than one *cluster* site from those available. Thus
ϕ is a function of the stacking order of the clusters. The
same problem would arise in a system with a high concentration
of noninteracting di-interstitials or noninteracting di-vacancies.
One must calculate ϕ as an ensemble average and the problem
is equivalent in complexity to calculating the configurational
entropy of the above system. A Monte Carlo calculation of ϕ
has been made for several Willis cluster models in UO_{2+x}
(Murch 1975b). The Mayer cluster theory, which has shown such
promise in the calculation of the configurational entropy of
similar models (Mitra and Allnatt 1979), would seem to be a
likely candidate for an analytical method of evaluation of ϕ.

As we noted in section 2.2.2.1 nonstoichiometry in compounds
such as UO_{2+x} and $Fe_{1-\delta}O$ is probably not accomodated by a single

type of cluster. *Ab initio* calculations by Catlow (1977) and Catlow and Fender (1975) suggest that a range of defect complexes is stable but each of the complexes is based on a common building block e.g., the $V_o - O_i^{('')} - O_i^{(')}$ defect kernel in UO_{2+x}. It would then be expected that such clusters continually fragment and rebuild in a manner which depends on the temperature and composition. For the purposes of establishing *the* diffusion mechanism such a picture is surely disconcerting. Firstly, it introduces a most nebulous nature for the transport mechanism and, secondly, any correspondence between mobile defect concentration and composition is entirely removed. Probably the only source of theoretical progress in the former problem is a direct Molecular Dynamics (MD) simulation; a start has, in fact, been made in the MD simulation of CaF_2 (Rahman 1976). To overcome the latter problem,. three approaches have been suggested. The first two, which involve an equilibrium between mobile and immobile defects will be dealt with below, the third, which involves the inclusion of interactions between defects, will be dealt with in section 3.2.3.

We first consider an equilibrium between mobile point defects and defect complexes. Norris (1977), in a theoretical study of thermomigration in $U_{1-y}Pu_yO_{2-x}$, suggested a quasi-chemical equation of the form

$$x_o = f_j(x_j),$$
(3.38)

where x_o is the mobile vacancy concentration and x_j is the equivalent vacancy concentration accomodated within structures of type j. The compositional deviation x is given by

$$x = x_o + \sum_j x_j$$
(3.39)

and in many cases of practical interest $x_o \ll \Sigma_j x_j$. In the particular

application $U_{1-y}Pu_yO_{2-x}$ Norris wrote the reaction

$$V + 2PuO_2 \rightleftharpoons (Pu_2O_3)_j , \qquad (3.40)$$

where V symbolizes a vacancy *not* associated with a pair of Pu

cations and j refers to the fact that not all Pu sites are

equivalent but depend on the proximity of other Pu_2O_3-like

complexes. The quasi-chemical equation governing the equilibrium

3.40 is

$$x_o = [x_j/(y_j - 2x_j)^2] \exp (-E_{B_j}/kT) , \qquad (3.41)$$

where y_j is the fraction of cations that are Pu ions able to

participate in the formation of complex j and where $x_o \ll x_j$ and E_{B_j}

is the binding enthalpy for eqn 3.40. The tracer diffusivity

D^* will be proportional to x_o of eqn 3.41.

In the second approach, Mitra and Allnatt (1979) did not

conceive of an equilibrium between point defects and defect

clusters, but rather an equilibrium between mobile *clusters* and

aggregates of clusters. As we noted above, in view of the strong

driving force for local reconstruction as distinct from isolated

point defects, this view would seem to be more realistic.

For UO_{2+x} Mitra and Allnatt (1979) wrote equations like

$$3m \rightleftharpoons t , \qquad (3.42)$$

where m and t symbolize monomer and trimer defect clusters.

Either species could be considered to ionize to give free holes.

So far, Mitra and Allnatt have considered only the effects of such equilibria on the partial molar enthalpy and entropy. Calculations have also been made with the inclusion of Debye - Hückel corrections although such corrections are hardly likely to be valid even at moderate defect concentrations (see section 3.2.3).

Further work along the lines of these approaches is likely to be quite fruitful and is enthusiastically encouraged by the author. Finally, we turn to the third approach which has received by far the greatest amount of attention.

3.2.3 Interacting Defects

The direct result of the introduction of interactions between defects (or ions) is the partitioning of defects into various levels of mobility. That is to say, there is a spectrum of mobility, ranging from free and probably highly mobile defects to virtually immobilized defects in ordered domains. It is important to note that the ordered defect domains may themselves move, albeit in an amoeboid manner, as mobile defects are added to and subtracted from the domains. The domains will, however, definitly be immobile in certain mixed systems e.g., CaO/ZrO_2, Y_2O_3/CeO_2 where the lower valent cations, which *may* be building blocks of the domains, are themselves immobile at temperatures where the high mobility of the anion vacancies is of interest. The ordered domains may, in actuality, be locally reconstructed defect complexes (see section 2.3) but such a level of sophistication has not been handled in the statistical mechanical approaches which have, almost exclusively, dealt with interactions between *point* defects.

More than 20 years ago, Lidiard (1956) suggested the application
of the Debye-Hückel theory of electrolyte solutions to account for
deviations from ideal behavior. Because of the much smaller
dielectric constant in ionic crystals than in water, the defect
concentration limit of the Debye-Hückel theory in ionic crystals
is less than 10^{-6} as opposed to about the equivalent of 10^{-4} in
water. A concentration of 10^{-6} is exceeded in the stoichiometric
alkali halides at temperatures greater than 500°C. The theory is
therefore quite unsuitable for highly defective solids. Nonetheless,
the *concept* alone of an effective dielectric constant has been
used with significant success in the statistical treatments of
the thermodynamics of highly nonstoichiometric ionic oxides such as
CeO_{2-x}, PuO_{2-x}, and UO_{2+x} (Atlas 1968a, b, 1970, Manes and Manes-
Pozzi 1976, Murch 1973, 1980b) see section 2.2.1. The extension
of that concept to diffusion in the above solids is yet to be
realized.

We start by considering the effect of both repulsive and
attractive nearest neighbour interionic interactions on the
activation energy for diffusion. In fig. 3.6 we have schematically
presented the result of such interactions. It is seen that
repulsive interactions decrease the barrier height for an isolated
atom in incremental factors of the interaction energy ε.* In an
analogous manner, attractive interactions increase the barrier
height. It is implied, of course, that the interaction
energy at the saddle point is zero. This may be justified, at
least in interstitial solid solutions, by appealing to the
theoretical result that indirect interactions oscillate in
sign with distance away from the given jumping atom.

---

*To preserve continuity with the literature we have used the symbol
ε rather than E_{ii} of section 2, $E_{ii} = -\varepsilon$.

ATTRACTIVE ATOM-ATOM INTERACTIONS

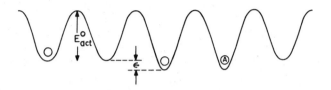

REPULSIVE ATOM-ATOM INTERACTIONS

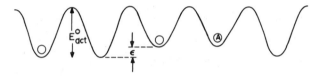

<u>Fig. 3.6</u>.   Sketch of the influence of nearest neighbour attractive and repulsive interactions on the activation energy for diffusion.

We now discuss the various statistical mechanical models which have quantified the above model in terms of all the relevant self-diffusion parameters. Surprisingly, the Bragg-Williams (BW) approximation has never been used for this problem. Because we believe this approximation to be useful in this respect we have sketched below what we believe is a formalism consistent with, and in the spirit of, the BW approximation.

By analogy with the statistical thermodynamic treatment at the BW approximation (see section 2.2.2.1) the configurational energy for defects *in a position to jump* would be $(Z - 1)\epsilon N_d / \gamma B$ where Z is the coordination number, N_d is the number of defects, B is the number of regular sites and γ is the ratio of interstices to regular sites[†]. The term $(Z - 1)$ arises because the defect must, of course,

[†]The inclusion of γ is only meaningful in the case of interstitials and the interstitialcy mechanism.

be next to an atom to effect an atomic jump. Defects completely surrou
by other defects are removed from consideration for diffusion. They do
of course, contribute to the *total* configurational energy.

Accordingly, we find that the activation energy for migration
in the BW approximation is given by

$$E_{act} = E_{act}^{o} + (Z - 1)\theta_d \epsilon , \qquad (3.43a)$$

where E_{act}^{o} is the activation energy for migration of an isolated
defect, $\theta_d = N_d/\gamma B$. Eqn 3.43 may now be substituted into eqns
3.5, 3.6 depending on the defect type. In the spirit of the BW
approximation wherein random mixing is assumed despite interactions,
the defect concentration (availability) remains as given in eqns 3.5 and
3.6. Finally, the tracer correlation factor may be approximated by
eqn 3.7 in the case of interstitial and vacancy diffusion. For
the interstitialcy mechanism, calculations show that the
correlation factor remains sensibly constant at least for
concentrations up to 20% (see section 3.2.1).

This BW approach would seem to be most appropriate for
diffusion in highly defective *nonionic* systems rather close to
their high temperature limit and for which the partial molar
thermodynamic quantities are well described by the BW approximation,
that is to say, the partial molar entropy is ideal and the partial
molar enthalpy is a linear function of composition (see section 2.3).
Examples are light atom diffusion in metals and carbides e.g.,
N and C in Fe, carbon diffusion in UC_x and ThC_x. When the stoichiometr
composition is of interest one may apply eqn 3.43 to both types
of defect e.g., anion vacancies and interstitials and substitute
directly into eqn 3.11. An example of such an application might
be carbon diffusion in hypo and hyperstoichiometric UC.

Recently, Murch (1980a) studied the fate of the defect interaction terms in $\ln D^*/\ln p_{O_2}$. Taking as the example the reaction

$$V_i + \tfrac{1}{2}O_2 \rightleftarrows O_i \, , \tag{3.43b}$$

mass action analysis leads to the fact that $[O_i] \propto p_{O_2}^{\frac{1}{2}}$, and, accordingly, from eqn 3.6, for small values of $[O_i]$ one finds that $\ln D^*/\ln p_{O_2}$ has a slope of $\tfrac{1}{2}$. At the BW approximation, from eqns 2.15 and 3.43 we find that[+]

$$\frac{d\ln D^*}{d\ln p_{O_2}} = \frac{1 + \theta_i E_{ii}(Z-1)/kT}{2(1 + \theta_i E_{ii} Z/kT)} \, . \tag{3.43c}$$

For typically encountered values of E_{ii} the essential cancellation of the factors containing E_{ii} leads again to the factor of $\tfrac{1}{2}$. In effect because of this cancellation the physics associated with non-ideality, as generated by the interaction terms, is *concealed* in a $\ln D^*/\ln p_{O_2}$ plot. Even if we adopt the more accurate approximations such as the Bethe for the chemical potential and the pair approximation of the Path Probability Method for D*, the same kind of cancellation occurs. The point to be made is that although there are considerably greater difficulties in experimentation, studies of the dependence of D* on *composition* are ultimately more rewarding in the understanding of the diffusive process.

In a most important contribution to the understanding of tracer diffusion and conductivity in highly defective solids, Sato and Kikuchi (1971 a, b) applied their Path Probability (PP) method to models of β and β'' alumina. The PP method is a time dependent version of the cluster variation method (Kikuchi 1951). The PP method is discussed at length in section 3.5.2. Very briefly, the method does not follow a specific migrating atom as is done in the

[+]The interstitial interaction energy, $E_{ii} = -\varepsilon$.

random walk approach, but rather, statistically treats the *entire*
system. The diffusion coefficient is calculated by imposing
a concentration gradient on the system and then evaluating the
ratio of the net flux of the diffusing species and the concentration
gradient of that species.

The β''-alumina system was modelled at the level of the *pair*
approximation of the cluster variation method. This approximation
is equivalent to the Bethe/Quasi-Chemical approximations in an
equilibrium treatment. Incidentally, the 'point' approximation
corresponds to the BW approximation. Only nearest neighbour *interionic*
interactions (see fig 3.6) were treated although more extended
interactions can be handled at the expense of much algebra.

In the case of a unit migration process, i.e., 'vacancy'
diffusion, and an interstitial concentration, $\rho\,(=1 - \theta_v)$, varying
from zero to unity, Sato and Kikuchi (1971 a, b) showed that it
was possible to expand the atomic jump frequency, Γ, in what is
actually a quite general form (c.f. eqn 3.5a)

$$\Gamma = VWZ\nu \, \exp(-G^o_{act}/kT), \qquad (3.44)$$

where V is termed the vacancy availability factor and represents
the probability that a vacant site exists at a nearest neighbour
site to an ion. As such, V is a *generalized* vacancy 'concentration'
term in a situation where the defects are nonrandomly distributed. It
may be thought of as depicting the effective vacancy concentration.
In the special case of no interactions and also the BW approximation

$$V = \theta_v , \qquad (3.45a)$$

and

$$= 1 - \rho. \qquad (3.45b)$$

W, in eqn 3.44 has been termed an effective frequency factor and contains the contributions to the jump frequency of the

interactions of the jumping ion with the surrounding ions.

In the non-interacting case W is trivially equal to unity,

but at the BW approximation W is given, for *interionic* interactions

(see eqn 3.43)

$$W = \exp \left[-\rho (Z - 1) \varepsilon/kT\right]. \qquad (3.46)$$

At the pair-approximation of the PP method in the honeycomb lattice,

V and W are given by

$$V = 2(1 - \rho)/(R + 1), \qquad (3.47)$$

and

$$W = [2(1 - \rho)/(R + 1 - 2\rho)]^2, \qquad (3.48)$$

where

$$R = [1 + 4\rho (1 - \rho)(\exp(\varepsilon/kT) - 1)]^{\frac{1}{2}}. \qquad (3.49)$$

In figs 3.7 and 3.8 we give the result for V and W at several values

of kT/ε. In fig 3.7 it can readily be seen that attraction between

ions (ε>0[†]) decreases the probability of finding a vacancy next

to a given ion, conversely, repulsion between the ions increases

the probability of finding an adjacent vacancy. As we have shown

already in the BW approximation, for the repulsive case, W

increases with increasing concentration as the coordination of an

ion increases, conversely, W decreases in the attractive case.

The correlation factor, f, was also obtainable from the PP

calculation. With no interactions the PP method gives eqn 3.7 (see

section 3.2.1) for the functional dependence of f on composition.

This linear equation is not rigorous but it is a quite reasonable

approximation as verified by exact Monte Carlo calculations

(see also section 3.2.1). More interesting is the effect on f of

[†]We have used the Ising-like convention, ε>0: attraction, ε<0: repulsion.

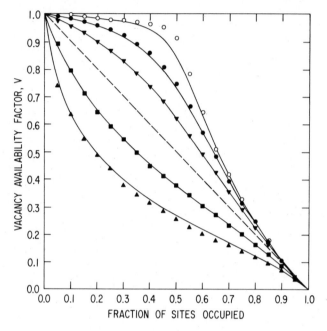

<u>Fig. 3.7</u>.   The vacancy availability factor, V, as a function of the
fraction of occupied sites at various values of kT/ε (honeycomb lattice
————: Path Probability Method.   ▲ (0.5), ■ (1.0),   ▼ (-1.0)
● (-0.5), O (-0.3):   Monte Carlo Method —— —— —— non-interacting i

temperature and sign of the nearest neighbour interaction energy.
The PP results are given in figs. 3.9 and 3.10. For increasing
attraction we see that the correlation factor *increases*. This
does not always happen in fact. Monte Carlo calculations by
Murch and Thorn (1977a) verified the above result in the honeycomb
lattice but the *converse* occurs in the simple cubic lattice (Murch
and Thorn 1977c). The precise physics of the situation remain
unclear, however. We may *interpret* the honeycomb result in the
following way. Because attraction between the ions causes the
vacancies to be repelled from the ions, after a vacancy/ion exchange,

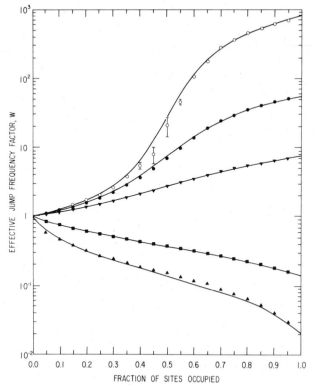

Fig. 3.8. The effective jump frequency factor, W, as a function of the fraction of occupied sites at various values of kT/ε (honeycomb lattice). ————: Path Probability Method. ▲ (0.5), ■ (1.0), ▼ (-1.0), ● (-0.5), O (-0.3): Monte Carlo Method.

the vacancy may be cooperatively pushed away by the surrounding ions thereby decreasing the probability of the reverse jump and increasing the correlation factor. In the simple cubic lattice, perhaps because of the increased potential *coordination* for nearest-neighbour ions, the inevitable islands of ions, which signal the onset of critical behavior, are more important in determining correlation effects than in the honeycomb structure. In such islands the correlation factor might be expected to be

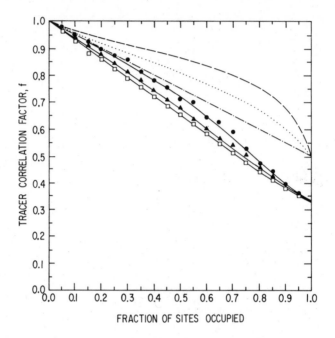

Fig. 3.9. The tracer correlation factor, f, as a function of the fraction of occupied sites at various values of kT/ε - *attraction only* (honeycomb lattice). — — — (0.5), ····(1.0), —·—·—·— (noninteracting): Path Probability Method. ●(0.5), ▲(1.0), □ (noninteracting): Monte Carlo Method.

at least reminiscent of f_0, the correlation factor of a filled lattice with a single vacancy. As the size of the islands increases, at lower temperatures, the correlation factor would then ultimately decrease.

In fig. 3.10 for the case of nearest neighbour repulsion between ions the correlation factor decreases with increasing repulsion. This seems to be a general result. It may be interpreted in the following way. Because repulsion between the ions causes the vacancies to be attracted to the ions, after

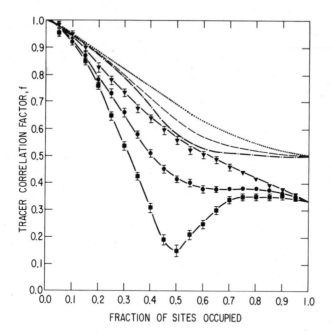

Fig. 3.10. The tracer correlation factor, f, as a function of the fraction of occupied istes at various values of kT/ε - *repulsion only* (honeycomb lattice). ——·——·—— · (-0.3), —— —— —— (-0.5), ····· (-1.0): Path Probability Method. ■(-0.3), ●(-0.5), ▼(-1.0): Monte Carlo Method.

a vacancy/ion exchange the vacancy is *retained* in the immediate vicinity thereby enhancing the probability of a correlated reverse jump and leading to a decrease in the correlation factor.

In a series of papers, Murch and Thorn (1977a-e) described the development of a Monte Carlo (MC) method to handle the problem of diffusion in a system of interacting ions/defects. Details of the method can be found in section 3.5.3. Very briefly, starting with the central idea of the random walk formalism (LeClaire 1970), wherein the migration of tagged atoms is traced, the MC method permits the atomic jump frequency to vary from site to site according to the local environment. The atomic jump frequency varies through both the interactions with other ions (see fig 3.6) and also the density of ions. These effects are not, of course, mutually exclusive. In the MC method a statistically averaged diffusion coefficient is given by calculating the distance a given tagged atom moves away from its origin in unit time. A MC calculation may also be performed in such a way that the diffusion coefficient is calculated from the ratio of the flux to concentration gradient (Murch and Thorn 1979a) but this method has not been developed extensively.

In both the Path Probability method and the Monte Carlo method there is a degree of arbitrariness in the way the tracer diffusion coefficient is decomposed into its component factors and, in particular, into factors like V, W and f. This would be true, in fact, of any theoretical treatment of diffusion in disordered solids. The diffusion coefficient itself, being a proportionality constant between flux and concentration gradient must remain well defined. For convenience and continuity most of the MC results have been compared with and expressed in terms of V, W and f. As an example we use β''-alumina. V, W and f as calculated by MC (Murch and Thorn 1977a) are shown in figs 3.7, 3.8, 3.9 and 3.10. The agreement in the case of V and W is generally very good with some deterioration in fit at and below the transition temperatures. This is to be expected as a result of the known overestimation of the transition point by the pair-approximation. In the case of f, the discrepancies are seen to be quite large with even an incorrect terminal value of f, i.e., f_o. Sato (private communication with the author), Gschwend, Sato and Kikuchi (1977), Sato and Kikuchi (1977) have indicated that these discrepancies can be traced back to an inappropriate

averaging of the flux. The corrected results for f agree with the
MC results as closely as do those for V and W.

With respect to the MC results for f, for $\varepsilon > 0$ (attraction)
little more can be said than that outlined above concerning the PP method.
For the repulsion side we note, however, that in actuality, a bold minimum
develops which is directly associated with the ordered phase
based on alternate occupation of sites by the ions. The ordered
phase is centered about 0.5 and its most likely boundaries have
been delineated in a renormalization group calculation by
Subbaswamy and Mahan (1976). In the ordered phase diffusion is
characterized by *disordering* jumps, as an ion jumps from a
'right' site to a 'wrong' site being followed by correlated
reordering jumps in the reverse direction. Such a situation must
lead to very low values of f. Interestingly, the minimum does
not occur exactly at 0.5 but is displaced to lower densities.
This may be traced back to the fact that at T>0 the entropically
most favoured ordered configuration does not occur at 0.5.

The effects of ion or defect interactions on the tracer
correlation factor have now been collectively called *physical
correlation effects* (Sato and Kikuchi 1975). The origin of such
correlation must be carefully contrasted to that of *geometrical
correlation effects*. The latter is the familiar 'classical'
correlation effect discovered originally by Bardeen and Herring
(1952). The geometrical effect arises as a consequence of following
a *distinguishable* atom which interacts with a defect.[†] The complete
generalization of *physical correlation effects* is yet to be
achieved but a sufficient condition seems to be a situation wherein
the atom jump frequency along a given jump vector is not equal
to the jump frequency in the reverse direction. As such, the
phenomenon of physical correlation is hardly likely to be

[†]
 The reader is referred to the review by LeClaire (1970).

confined to highly defective solids, indeed it can be expected to occur in partially ordered alloys as well, but this avenue has not been explored *with physical correlation in mind*. The phenomenon of physical correlation will be dealt with further in section 3.3 when we come to discuss the physical correlation factor.

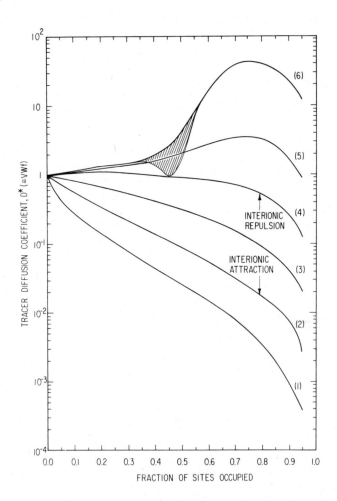

Fig. 3.11.   The tracer diffusion coefficient D* (=VWf) as a function of the fraction of occupied sites at various values of kT/ε.   Curves 1-6 correspond to kT/ε = 0.5, 1.0, ∞, -1.0, -0.5 and -0.3 respectively

It is now possible to assess the dependence of the tracer diffusion coefficient on composition by taking the product of V, W and f. We exemplify the discussion with the results of the MC calculations of Murch and Thorn (1978b) in the honeycomb lattice (fig 3.11). Examination of this figure reveals that attraction between the ions leads to suppressed diffusion compared with non-interaction. Conversely, repulsion leads to enhanced diffusion except within the ordered region. It is interesting to note that a model which provides for nearest neighbour repulsive interactions between the ions is sufficient to explain the high diffusivity in real β''-alumina (conduction plane density about 0.7). This is a direct result of the combined effects of an increased accessibility of ions to neighbouring vacant sites and a decrease, through interactions, of the atomic migration energy.

An important result of the PP and MC calculations is that the temperature dependence of the many-body terms VWF *as a whole* is such that it gives an activation energy (positive or negative) in addition to E^o_{act}, the activation energy of an *isolated* ion. Contrast this with the BW approximation where only W was concerned with an additional temperature dependence (eqn 3.46). The model for β''-alumina where interacting ions are distributed over a lattice of equivalent sites *a priori* is useful to examine the effect of the order/disorder transition on diffusion (Sato and Kikuchi 1977). The calculations predict a sharp break in the log D vs 1/T curve (fig 3.12) at the order/disorder temperature. The sharp break is a result of a discontinuous jump in the activation energy provided by VWf. But only a minor part of the decrease in D by ordering is the increase in the activation energy of ionic motion caused by ordering, i.e., VW. The major part of the decrease in D is due to the decrease in f upon ordering.

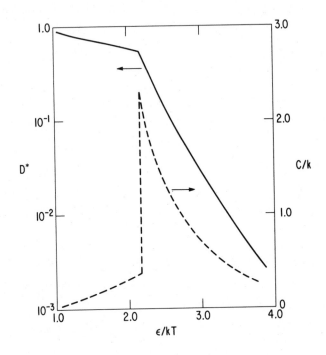

Fig. 3.12.   Sketch of the dependence of both D* (VWf) and specific heat C on temperature. The break in D* occurs at the order/disorder temperature as also shown by the λ-type anomaly in C.

A similar model involving interactions between the diffusing atoms has since been used in a MC calculation to describe light atom interstitial diffusion in metals (Murch and Thorn 1977c) and specifically tracer carbon diffusion in γ Fe (austenite) (Murch and Thorn 1979b). Quite reasonable agreement with experimental data (Parris and McLellan 1976) could be obtained assuming nearest neighbour repulsion between carbon atoms. The magnitude of the interaction energy was obtained from a previous MC study

of the carbon activity (Murch and Thorn 1979e) see section 2.2.2.1.
Better agreement could have been obtained if E^o_{act} was allowed to
vary slightly with composition, an inherently reasonable assumption
in view of the lattice parameter change, but this is beyond the
confines of the lattice gas treatment. Application to other
interstitial systems which exhibit wide compositional homogeneity
e.g., N in Fe, H in Ta, Nb, V and O in Zr has yet to be realized.
Murch and Thorn (1977b) also used a similar model to describe
carbon diffusion across the compositional range $1.0 < x < 2.0$ in UC_x.
This work was an evolution of the non-interaction model used in
the same system, see section 3.2.1 and also Murch (1975a). As
implied by the compositional form of the carbon activity, $a_c(x)$
(Murch and Thorn 1976), nearest neighbour *attraction* between
C_2 groups was assumed. While the compositional dependence of D^*
on x was only in fair agreement with that observed experimentally,
the form of $E_{act}(x)$ was in quite reasonable agreement with that
observed.

Sato and Kikuchi (1971 a, b) have also applied the PP method to
β-alumina. In this case the conduction plane was conceived as
a superposition of two equivalent triangular sublattices A and B
separated by an energy gap ω. In a more general sense, we may classify
the low energy sites as *normal* lattice sites, the high energy sites may
then be termed *interstices*. The ions are partitioned between these
sublattices subject to both ω and the interaction energy ε. For the
assumed vacancy mechanism the total W is now given by the harmonic mean

$$W^{-1} = \tfrac{1}{2}(W_A^{-1} + W_B^{-1}) \tag{3.50}$$

whereas the total V is given by the weighted mean

$$V = (2\rho)^{-1} (pV_A + qV_B) \qquad\qquad (3.51)$$

where p and q are the densities of the ions on the sublattices
A and B respectively. Sato and Kikuchi also expressed the total
f as an average of f_A and f_B but it is doubtful whether this is
valid in a completely rigorous treatment. The PP results for
the compositional dependence of V, W and f and their MC
counterparts are similar for β-alumina and we refer to Murch and
Thorn (1977d) for details. An important consequence of the model
is the emergence of an important interstitialcy-like character to the
diffusion process and we shall disuss this further in section 3.3.

The temperature dependence of VWf is of some interest in
connection with the Faraday transition (O'Keeffe 1976). The
Faraday transition, which is well documented for crystals with
the fluorite structure (O'Keeffe and Hyde 1976), is characterized
by a smooth change in the diffusivity and a specific heat curve
with a round peak of the forms shown in fig 3.13. Such a
transition bears some resemblance to melting and the term
'sublattice melting' is sometimes used. A transition with the
same characteristics is generated in the β-alumina model by a
cooperative excitation of ions from the low energy sublattice
(A) to the high energy lattice of 'interstitial' sites B. We
use the term *cooperative* to emphasize the fact that *interactions*
among the ions create a more abrupt transition than ordinary Frenkel-
pair formation. The fluorite structure could, of course, itself be
modelled with the PP method but this has not yet been attempted.
Such a calculation would be most appealing because it provides a
formalism in which the Frenkel defect formation process is inextricably
linked to the diffusion process itself.

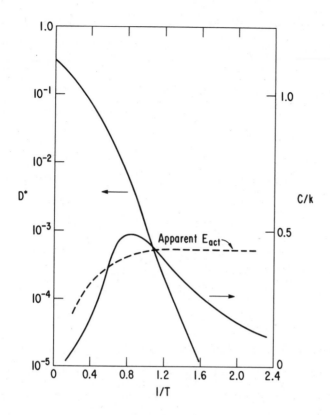

<u>Fig. 3.13.</u> Change of the tracer diffusion coefficient, D*, and the specific heat, C and apparent activation energy which accompany a cooperative excitation of diffusing ions to interstitial sites (Faraday Transition).

It is clear that the magnitude and sign of ε can be deduced from compositional and temperature plots of D*. The realization of this is quite recent and accordingly, few studies have been made. We refer to the following analyses: O diffusion in UO_{2+x} (Murch 1975b), C diffusion in UC_x (Murch and Thorn 1977b) and C diffusion in γFe (Murch and Thorn

1979b, Parris and McLellan 1976). It is also worth pointing out
that in situations where essentially only one defect is responsible
for diffusion, then the compositional dependence of the heat of
solution should roughly parallel that of the activation energy of
diffusion, see eqns 2.15 and 3.43a and also Murch (1979,1980d).
Correlations of this type may be useful in *predicting* the likely
course of a compositional dependence. There are rather few systems
where the requisite data are available to illustrate the hypothesis.
In the UC_x (1.0 < x < 2.0) system, within the reported uncertainty,
the partial molar enthalpy of carbon and the activation energy of
carbon diffusion do show a similar dependence on composition, see
Murch and Thorn (1976,1977b).

3.2.4 The Isotope Effect

The isotope effect remains, with the Haven ratio (see section 3.3)
a time-honoured means of identifying the diffusion mechanism through
the phenomenon of correlation. The isotope effect in diffusion has
been expressed (Schoen 1958 and Tharmalingham and Lidiard 1959)

$$\frac{1 - (D_\beta/D_\alpha)}{1 - (\Gamma_\beta/\Gamma_\alpha)} = f \, , \qquad (3.52)$$

where α and β refer to two isotopes of mass m_α and m_β respectively.
If the motion of the migrating atom is coupled with the surroundings,
the theory of classical statistical mechanics leads to the
following expression for $\Gamma_\beta/\Gamma_\alpha$

$$1 - (\Gamma_\beta/\Gamma_\alpha) = \Delta K [1 - \left(\frac{m_\alpha}{m_\beta}\right)^{\frac{1}{2}}] \, , \qquad (3.53)$$

where ΔK is the *fraction* of the total translational kinetic energy at the saddle point, which is associated with motion in the direction of the jump, which actually resides in the hopping atom. Eqns 3.52 and 3.53 lead to the following general expression for the isotope effect

$$\frac{1 - (D_\beta/D_\alpha)}{1 - (m_\alpha/m_\beta)^{\frac{1}{2}}} = f\Delta K. \qquad (3.54)$$

For diffusion mechanisms that involve more than one atom, Vineyard (1957) deduced that the quantity $(m_\alpha/m_\beta)^{\frac{1}{2}}$ should be replaced by

$$\left\{ \frac{(n-1)m + m_\alpha}{(n-1)m + m_\beta} \right\}^{\frac{1}{2}}$$

where n is the number of atoms which participate in the jump process and m is the average mass of the nontracer atoms. Equations analogous to eqn 3.53 can also be derived for cases where there are several jump frequencies or mechanisms.

Noting eqn 3.53 one can determine $f\Delta K$ by measuring the relative diffusion coefficients of two isotopes α and β. This is achieved, incidentally, without the need for distance and time to be known explicitly, by diffusing two isotopes simultaneously from a very thin source. The following equation is then appropriate

$$\ln(c_\alpha/c_\beta) = \text{const} - [1 - (D_\beta/D_\alpha)]\ln c_\beta , \qquad (3.55)$$

where c is the specific activity.

There have been very few reports of measurements of fΔK in highly defective solids. Although not really qualifying as highly defective solids, NiO and CoO are of interest here. Both these almost-stoichiometric oxides can exhibit cation deficiency. Reported values of fΔK were. consistent with cation diffusion via a single vacancy mechanism (f = 0.7815) with ΔK = 0.78 and 0.75 for NiO and CoO respectively (Volpe, et al., 1971 and Chen et al., 1969). Recently there has been a careful study of fΔK at three temperatures and as a function of composition in $Fe_{1-\delta}O$ (Chen and Peterson 1975, 1976). These results are displayed in fig 3.14. It can be seen that fΔK decreases quite strongly with composition, but any temperature dependence is lost in the statistical uncertainty.

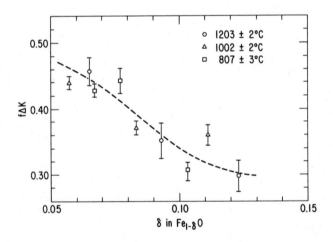

Fig. 3.14. The values of fΔK versus the deviation from stoichiometry for iron self diffusion in $Fe_{1-\delta}O$ (after Chen and Peterson (1975)).

Quite similar behaviour was found for f alone by Murch (1973), deBruin and Murch (1973) and Murch (1980b) in their Monte Carlo studies of tracer correlation in the lattice gas with nearest neighbour vacancy blocking i.e., vacancies cannot be adjacent. The decrease of f rather than an increase toward unity may be ascribed to the physical correlation effect described in section 3.2.3. But as we have already discussed in section 3.2.2 the defects responsible for nonstoichiometry in $Fe_{1-\delta}O$ are not, in actuality, point defects but coherent defect clusters (Koch and Cohen 1969) and diffusion probably occurs via concerted motion of the clusters. Nonetheless, the physical correlation effect which results from hard vacancy - vacancy interactions may well be reminiscent of hard cluster - cluster interactions as it is in the analogous case of interstitials and interstitial clusters in UO_{2+x} (Murch 1975b). That there are strong repulsive interactions between defects, probably clusters in fact, in $Fe_{1-\delta}O$ is also implied by Libowitz's (1968) calculations of the configurational entropy, see section 2.2.2. We should also recognize that a correlation factor greater than unity is also a possibility (see section 3.2.2) but indirect results for f via the modified Darken equation do not support this (see section 3.4.2).

Both the value of ΔK and its variation with composition are completely unknown in highly defective solids. A Molecular Dynamics computer simulation along the lines suggested by Bennett (1975) would seem to be the most promising line of attack.

In many highly defective solids it is the anion sublattice which exhibits the disorder and this poses a most difficult

problem in the acquisition of two suitable tracer species. Some
of the recent non-radioisotope techniques have been reviewed in
Appendix II. To the author's knowledge none of these techniques
has ever been used for the isotope effect except for the prompt
γ analysis of oxygen isotopes in UO_{2+x} (Murch 1973). The nuclear
reactions chosen in that study

$$^{17}O(p,p'\gamma)^{17}O$$

$$^{18}O(p,\gamma)^{19}F$$

$$^{16}O(p,\gamma)^{17}F$$

can in fact be induced simultaneously with a proton energy of
about 1.940 MeV and the resultant γ rays specific to ^{16}O, ^{17}O
and ^{18}O recorded (Russell and Murch 1973, Bird, Russell and Murch
1974). Results from UO_{2+x} were, however, inconclusive because
the γ count rates were not high enough at proton currents low
enough to avoid thermal damage to the sample. In the future,
when more efficient γ detectors with the requisite high resolution
have been developed, the method will undoubtedly prove viable.
We refer also to the possible use of ion beam mass spectrometry,
see appendix II.

3.3 Ionic Conductivity

3.3.1 Types of Highly Defective Solid with Ionic Conductivity

Compounds which exhibit prodominantly ionic conductivity
will be those compounds which possess wide band gaps to ensure
very low intrinsic electronic and hole mobilities. Thus pure
nonstoichiometric compounds which exhibit large deviations from

stoichiometry as a result of a favoured valence change are *not*, in general, ionic conductors. Contrast thus $UO_{2(+x)}$ which exhibits a wide homogeneity range and is a p-type semiconductor, to ThO_2 which is essentially stoichiometric and a predominantly ionic conductor.

A few compounds e.g., AgI, exhibit a high degree of lattice disorder with such a low migration energy of the charge carriers that they are ionic conductors despite a small band gap. In other compounds the electronic conductivity can be minimized and the ionic conductivity maximized by the addition of an altervalent cation to the lattice. Examples are the solid electrolytes based on ZrO_2 and ThO_2 doped with CaO and Y_2O_3. The high ionic conductivity is due to the high concentration of charged anion vacancies produced as a result of electroneutrality on the addition of Ca^{2+}, Y^{3+} etc. Finally, we have compounds typified by the β-alumina family which probably are not nonstoichiometric at all, but are true 'stoichiometric' compounds whose composition depends on the thermal history and level of dopant. Again the high ionic conductivity is due to extensive lattice disorder and a low migration energy of the charge carriers.

These three types of highly defective solids which exhibit very high ionic conductivity are often called superionic conductors or fast ion conductors. The study of superionic conductors has developed rapidly in recent years mainly as a result of the technological importance of these solids as battery and fuel-cell electrolytes. The subject has been reviewed extensively (Ingram and Vincent 1977, McGeehin and Hooper 1977, Hayes 1978, Hooper 1978, Funke 1976, Shahi 1977, Hagenmuller and van Gool 1978, Salamon 1979) and a number of major conferences have been held (van Gool 1973,

Mahan and Roth 1976, Vashishta, Mundy and Shenoy 1979). In the
present monograph it is not our intention to review all the
progress that has been made. We have chosen to confine our attention
to two subjects which we believe are usually inadequately covered: the
link between diffusion and ionic conductivity as expressed in the
Nernst-Einstein equation and physical correlation effects. This
approach provides a theme of continuity with sections 3.2 and 3.4
and fits in with our desire to take a statistical mechanical
viewpoint in this monograph.

3.3.2   The Nernst-Einstein Relation and the Haven Ratio

The macroscopically measured ionic conductivity, σ, is
conveniently defined at the atomistic level by

$$\sigma = \sum_i n_i q_i u_i. \tag{3.56}$$

Where u_i is the mobility of species i, q_i is the charge and n_i
is the number of charge carriers of species i per unit volume.
We shall henceforth confine ourselves to a single type of charge
carrier. The link between the diffusion coefficient, D_q, of the
charge carriers and the mobility is provided by the *very general*
Nernst-Einstein relation (Mott and Gurney 1954, Lidiard 1956)

$$\frac{u}{D_q} = \frac{q}{kT}. \tag{3.57}$$

By analogy with eqn 3.1 it is permissible to expand D_q in the
following way (LeClaire 1975):

$$D_q = \tfrac{1}{2} \lambda_q^2 \Gamma_q f_I \, , \tag{3.58}$$

where λ_q is the component of the jump distance in the diffusion direction, Γ_q is the charge carrier jump frequency and f_I is the correlation factor for the charge carriers. We have chosen to use the symbol f_I rather than f_q because of historical reasons and the fact that f_I is of a more general origin as we shall see later. The correlation factor f_I has variously been called the physical correlation factor, the conductivity correlation factor, the drift mobility factor, etc. The first of these is probably the most used.

In combination with eqns 3.1, 3.57 and 3.58 the Nernst-Einstein relation (eqn 3.57) becomes, in terms of D^* and σ

$$\frac{D^*}{\sigma} = \frac{kT\Gamma}{q^2 n \Gamma_q} \left(\frac{\lambda}{\lambda_q} \right)^2 \frac{f}{f_I} \, . \tag{3.59}$$

When the charge carriers are vacancies e.g., calcia stabilized zirconia, $\lambda = \lambda_q$ and $\eta\Gamma = n\Gamma_q$ where η is the number of ions per unit volume. From eqn 3.59 we find that

$$D^*/D_\sigma = f_v/f_I = H_R \, , \tag{3.60}$$

where D_σ, the conductivity diffusion coefficient is defined by:

$$D_\sigma = kT\sigma/\eta q^2 \, , \tag{3.61}$$

and f_v is the tracer correlation factor for vacancy diffusion.

The ratio of D^* to D_σ is commonly called the Haven Ratio, H_R.
In the case where ions on equivalent sites a priori are charge
carriers e.g., β''-alumina, the Haven Ratio is

$$H_R = f_i/f_I , \qquad (3.62)$$

here $D_\sigma = D_q$ since $\eta = n$

and f_i is the tracer correlation factor for interstitial diffusion.
Clearly, in the hypothetical situation often useful in modelling
highly defective solids where the entire composition range from
a single interstitial to a filled lattice can be traversed f_v
and f_i are points somewhere near the termini in a plot of the
correlation factor, versus composition, see sections 3.1.1 and
3.2.3. We will continue therefore to use the single symbol,
f, as the correlation factor for a unit migration process.

If the charge carriers are interstitials which move by
interstitialcy jumps e.g., AgBr (Weber and Friauf 1969), two
ions move for each defect jump and $\eta\Gamma = 2\Gamma_q n$. For a collinear
interstitialcy jump, the Haven Ratio is then given by

$$H_R = f_i^{cy}/2f_I , \qquad (3.63)$$

where f_i^{cy} is the tracer correlation factor for interstitialcy
diffusion; again a composition dependence of f_i^{cy} should be
recognized, see section 3.2.1.

For many years neither the compositional dependence of f
nor the very existence of f_I was recognized. The important
papers by Sato and Kikuchi (1971 a, b) were responsible for

changing this attitude. We have already dealt with the compositional dependence of the tracer correlation factor in sections 3.2.1 and 3.2.3. We now focus attention on f_I.

As implied in eqn 3.58, f_I is a correlation factor for *indistinguishable* particles, be they ions or defects.[†] The configurational factor or distinguishability factor which is such an essential part of the *tracer* correlation factor is *not* present, see section 3.2.3. The origin of f_I is in the physical correlation effect, itself a natural consequence of *nonideality*. *The physical correlation factor may only be calculated from a time dependent statistical mechanical treatment.* Thus far, only two methods have been advanced to calculate f_I. The first of these is the Path Probability (PP) method of Sato and Kikuchi (1971 a, b). The principles of this method have been briefly discussed in section 3.2.3 and more fully in section 3.5.2. To calculate f_I, Sato and Kikuchi imposed an electric field on the system and calculated the degree of nonequilibrium resulting from the field. The second method of calculation is the Monte Carlo (MC) method of Murch and Thorn (1977e). In this case the mobility is assumed to be proportional to the electric field and the constant of proportionality and hence f_I is calculated in an equilibrium sense. A description of the computational realization of this is described in section 3.5.3.

Originally, Sato and Kikuchi (1971 a, b) found a non-unity value of f_I only for the case of their site inequivalent model for β-alumina. In this, the two dimensional plane was conceived as

---

[†]It is *not* related to the pseudo-'tracer' correlation factor obtained by *tagging* defects.

the superposition of two lattices, separated, *a priori*, by an energy difference, ω. The ions were distributed over the sites subject to ω and a nearest neighbour interaction energy ε. The ions were assumed to migrate by the vacancy mechanism. A typical result for f_I is shown in fig 3.15. The minimum is the result of strong ordering at the composition $\rho = \frac{1}{2}$. At this composition a locally disordering jump i.e., a jump from a low energy site to a high energy site tends to be followed by a locally reordering jump in the reverse direction. At higher compositions enough of the high energy sites are occupied for an interstitialcy - like mechanism to operate. In this, an ion which leaves a low energy site to proceed to a high energy site probably will not reverse that jump because another ion on a neighbouring high energy site meanwhile moves into the site. Such a mechanism has, of course, some physical correlation associated with it but it does not seriously disrupt the local order. Although the two atoms do not move in unison this is immaterial in a statistical mechanical treatment which does not contain continuous time but rather deals in time *steps* to move from one configuration to another. The fact that this model, which is built on, ostensibly, vacancy diffusion, also contains the interstitialcy mechanism as a natural consequence of ordering, has never been fully appreciated.

Surprisingly, at the time, Sato and Kukuchi did not find values of f_I less than unity in the β''-alumina case where $\varepsilon \neq 0$ and where all sites are equivalent a priori. i.e., $\omega = 0$. In their series of MC computer simulations on β and β''-alumina, Murch and Thorn (1977 a,d,e, 1978 b) were able, firstly to verify the above results for β-alumina, and secondly to show that $f_I \leqslant 1$ in the simpler β''-alumina model. MC results for f_I above and

below the order/disorder temperature are shown in fig 3.16. It
is clear that only short range order is required for physical
correlation. Sato and Kikuchi (1977) and Gschwend, Sato and
Kikuchi (1977) have indicated that their result $f_I = 1$ could be
traced back to an inappropriate averaging procedure and now that
this has been resolved they find $f_I<1$ and that there is excellent
agreement between the two methods. Further MC work has generalized
physical correlation to the cases where there is site inequivalence
alone and also where there are obstacles for the diffusion of *non*
interacting ions (Murch and Thorn 1978c, Murch and Rothman 1980).

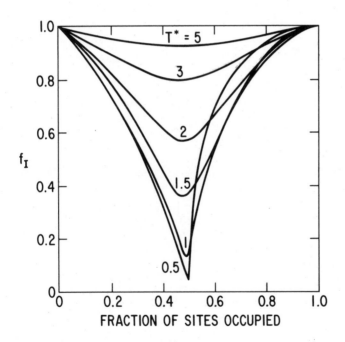

Fig. 3.15.   The physical correlation factor, f_I, in the honeycomb
model of βalumina with $\omega = 5|\varepsilon|$ (Sato and Kikuchi 1971a, b) as a
function of the fraction of occupied sites at various values of
reduced temperature $T*(=kT/\varepsilon)$.

We can now summarize these findings by observing that physical correlation is the result of 'ordering' among the ions. Such ordering may come about (1) simply through the apportioning of ions over two (or more) types of site or (2) through interactions between the ions, or (3) dodging of obstacles which are, say, randomly distributed. So far f_I has always been found to be less than or equal to unity. The possibility of f_I being greater than unity has not been pursued. All of our discussion above has been couched in the formalism of the vacancy mechanism although the same arguments are almost certain to pertain to the interstitialcy mechanism but work along such lines has not been essayed.

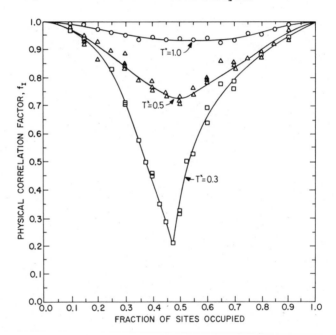

Fig. 3.16. The physical correlation factor, f_I in the honeycomb lattice as a function of the fraction of occupied sites at various values of reduced temperature $T^*(=kT/\varepsilon)$. □ (0.3), △ (0.5), O(1.0): Monte Carlo Method.

But is this physical correlation factor, f_I, a true correlation factor? Perhaps not. Calculations with the PP and MC methods show only that in eqn 3.59 a new factor, let's *label* it f_I, appears in the denominator. On comparing eqn 3.59 with the supposedly very general original form of the Nernst-Einstein relation, eqn 3.57, one finds that this factor has the *status* of a correlation factor, see its position in eqn 3.58. But suppose eqn 3.57 is not so very general and indeed does not apply to the situation of interacting ions. This would not detract from the importance of the presence of f_I in eqn 3.59 or the Darken equation (eqn 3.69) but f_I would have to be interpreted purely as a manifestation of non-ideality in the flow of the diffusing ions, i.e., a physical *flow* factor. It is instructive perhaps to look at the history of the vacancy-wind-effect (Manning, 1968). In the example case of impurities and vacancies (both at low concentrations) in a host when an electric field is applied, one can create a situation wherein vacancies approach an ion more frequently from one direction than the other. This phenomenon, known generally as the vacancy-wind-effect, changes the local vacancy distribution near an ion and may cause each ion to have an enhanced probability of jumping in a direction opposite to the vacancy flux. This leads to the addition of a factor in the place of f_I in the denominator of eqn 3.59. If we had let a computer probe this effect (Murch and Thorn 1979f) *without* the benefit of Manning's transparent theory we might have been led to introduce a spurious f_I which could be >1 and could even be negative! Thus it is prudent at this stage to be open-minded on the question of whether f_I is a *true* correlation factor in the case of interacting ions.

We turn now to the behaviour of the *Haven Ratio*, H_R, in the presence of interactions. Sato and Kikuchi (1971a, b) found that the strong temperature dependence of both f (see section 3.2.3) and f_I as a result of physical correlation effects tended to cancel in the ratio such that the ratio remained within the limits of unity and f_o, the value of f for a system with a single vacancy. Later, Murch and Thorn (1977e) showed that certainly above the order/disorder boundary in the model of β''-alumina, H_R behaved rather like f in the non-interacting case. The issue is clouded somewhat by the statistical uncertainty of the MC results which is, of course, amplified from that inherent in f and f_I alone.

For isothermal cuts through the ordered region in the model for β''-alumina, there are inflections at least, in H_R in the vicinity of the second order transitions at the entry to, and exit from, the ordered region Fig. 3.17a. This clue prompted a close MC examination of H_R in the simple cubic athermal lattice gas (Murch 1980b). In this model, which is the familiar nearest neighbour blocking model (O'Keeffe 1970), there is a single second order transition between the liquid-like distribution at low densities and the solid-like distribution at high densities. In this case, H_R goes through a minimum precisely at the point of the second order transition (fig 3.17b). Interestingly, in none of the above cases does either f or f_I in a compositional plot appear to exhibit an inflection at the transition point.

These findings would suggest that H_R may, in fact, be quite a sensitive probe of order/disorder transitions. So far, we are unaware of any measurement of H_R as a function of *composition* in a highly defective solid where the lattice gas model might reasonably be termed appropriate. Now in their study of Na-β-

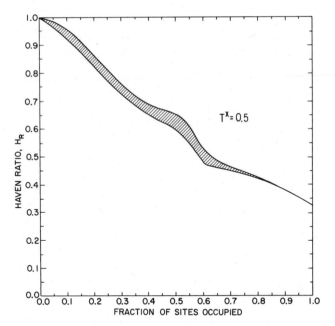

Fig. 3.17a. The Haven Ratio, H_R, as a function of concentration of interacting ions in the honeycomb lattice model of β"-alumina with $kT/|\varepsilon| = 0.5$.

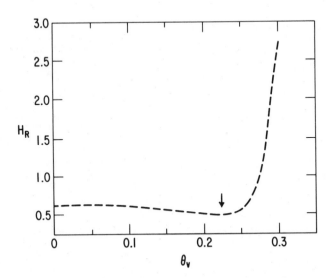

Fig. 3.17b. The Haven Ratio, H_R, as a function of vacancy concentration, θ_v, with nearest-neighbour vacancy blocking (simple cubic lattice).

alumina Kim, Mundy and Chen (1979) found that H_R exhibited a
break with decreasing temperature, fig. 3.17c, yet no abrupt
phase transition has been reported in that area. Disregarding
this break, Wolf (1979) modelled only the main trend of $H_R(T)$ on
the realistic basis of the occupation of three types of site in
the conduction plane (Roth et al., 1979), the interstitialcy
mechanism and the partial association of Na ions with the charge-
compensating oxygen interstitials. However, the number of parameters and the neglect of defect -wind-effects in the calculation
of H_R detracts somewhat from the physical significance of the
results obtained.

Sato and Gschwend (1980) have recently shown that the model
which was originally developed for β-alumina* and recently corrected
for deficiencies in f (see earlier this section), shows the break,
fig. 3.17d, in $H_R(T)$ as observed. This is a remarkable result.
They showed that the strong dependence on temperature of H_R is
related to the repopulation of conduction ions among the two sites,
to the tendency of creating preferred sites because of the
existence of ω and of repulsive interactions, and finally to the
percolation sensitivity of tracer diffusion. The last of these
produces the break in $H_R(T)$ because of a *smooth* transition and a
resultant 'percolation difficulty' of the tracer ions. Because
the ions always have a finite jump frequency and the connecting
network of ions can be made infinite, percolation never stops
completely and there is no true percolation threshold. The term
'percolation difficulty' was coined to refer to this situation.
This explanation is undoubtedly also the cause of the inflections
and turning points in H_R as shown in figs 3.17a,b; at the order/

---

*This model is by no means especially realistic according to current
structural data.

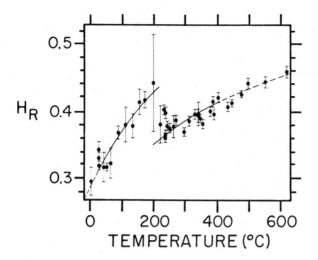

Fig. 3.17c. The Haven Ratio, H_R, as a function of temperature in Na-β-alumina (after Kim et al. (1979)).

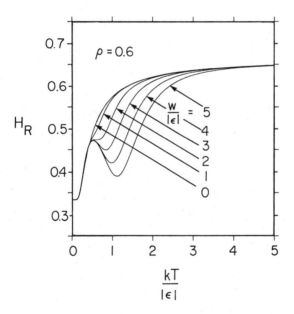

Fig. 3.17d. The Haven Ratio, H_R, as a function of temperature in the site inequivalent honeycomb lattice with site energy difference, ω, and interaction energy, ε, at a concentration of 0.6 of conducting ions (after Sato and Gschwend (1980)).

disorder transition one generates an abrupt change in the complete-ness of the network of ions and this changes the ability of the ions to percolate.

The important points to be made are that (1) the strange break in H_R is a natural consequence of a theory of many-body diffusion based on a simple model with a reasonable choice of a few parameters and (2) H_R contains a good deal of interesting physics associated with the phenomenon of percolation/diffusion.

Single values of H_R have been recorded in several superionic conductors. Whittingham and Huggins (1971 a, b) measured the ionic conductivity in Na and Ag β-alumina and compared them with the tracer diffusion coefficient measured by Yao and Kummer (1967). They found H_R to be 0.702 and 0.706 when corrected for the actual value of η, the number of change carriers per unit volume (Haven 1978) These values are not in good agreement with those found by Kim et al., (1979) nor with the value for H_R of 0.505 assuming the interstitialcy mechanism and no physical correlation (Haven 1978).

Values for H_R have also been reported in α-AgI by Jordon and Pochon (1957) Kvist and Tärneberg (1970) in α-Ag_2S by Okazaki (1967), Barthowicz and Mrowec (1972) and Allen and Moore (1959) in α-$RbAg_4I_5$ by Owens and Argue (1967), Bentle (1968), but no rigorous formulation has been advanced to explain the data, see Funke (1976).

Finally, in the group of oxides which exhibit a high oxygen conductivity, Simpson and Carter (1966) measured H_R in calcia stabilized zirconia. At the composition of maximum conductivity $Zr_{0.85}Ca_{0.15}O_{1.85}$ H_R was found to take a value of 0.65. Steele (1972) has expressed surprise that this value is close to that for the single vacancy diffusion in a simple cubic lattice (0.654) since he considered that cooperative mechanisms resulting from defect-defect interactions might be expected to give a rather different value.

As we pointed out earlier in this section the cooperative
effects as manifested in strong temperature and compositional changes
in f and f_I separately, tend to cancel in their ratio. In the case
of the nearest neighbour blocking model, which is a familiar model
for ionic conductivity in calcia stabilized zirconia, H_R does in
fact remain close to 0.654 for a wide range of composition, fig.
3.17 b(Murch 1980b). It should be pointed out, however, that a
considerable change in the activation energy for ionic conductivity
with composition has been reported (Carter and Roth 1968) and this
is, of course, beyond the capability of the blocking model. What
is required in such systems is a model which includes effective
association between the immobile lower valent cations and the anion
vacancies. Such a model must include physical correlation effects
for the rigorous analysis of the ionic conductivity.

In summary, statistical mechanical models based on the
vacancy mechanism suggest that $H_R \sim$ constant for very strong
interactions (blocking). Furthermore, H_R may be a probe for
phase transitions. We should also like to point out at this juncture
that H_R is not just accessible from measurements of the tracer
diffusivity and ionic conductivity i.e., via the Nernst-Einstein
relation. In section 3.4 it is established that H_R also occurs
in the Darken equation appropriate for highly defective solids
and this would provide a most valuable cross-check in systems
amenable to both experiments.

3.4 Chemical Diffusion

3.4.1 General Remarks

Chemical diffusion is the means by which a compositional change is accomodated in the solid state. It is clear that chemical diffusion is of particular interest in nonstoichiometric compounds. We have restricted our present discussion to compositional changes *within* a single phase field of a nonstoichiometric phase, although in the case of certain pseudo-nonstoichiometric phases such a restriction is not entirely meaningful.

A structural characteristic of many highly defective solids is an intermesh of a rigid sublattice with a low mobility and a highly defective sublattice with a high mobility. When a compositional change is imposed on the crystal it has, therefore, seemed appropriate to consider only atomic diffusion in the mobile sublattice. But such a restriction where the crystal is transformed to a pseudo-unary system is appropriate only to solids with a highly mobile electron concentration. Examples are the metallic-like interstitial solid solutions, carbides and nitrides. In ionic crystals, for electroneutrality reasons, one must have a concomitant accompaniment of the charged atomic defect by an oppositely charged electronic defect. This is the phenomenon of *ambipolar diffusion*. In the extreme case of a predominantly *electronic* conductor the rate of compositional change is still dependent on the atomic mobility; conversely, in the case of a predominantly *ionic* conductor, the rate of compositional change is limited by the electronic defect mobility, or, possibly, by the concentration of scarce, but very mobile, electronic defects.

Most of the theoretical understanding of chemical diffusion
in highly defective solids pertains to chemical diffusion in
solids with high electronic conductivity and we shall concentrate
mainly on this in the next sections. Subsequent sections will
deal with the special problems of chemical diffusion in the
presence of stress and chemical diffusion in solids exhibiting
crystallographic shear.

3.4.2 The Darken Equation

In a highly defective solid with a single mobile sublattice
the chemical diffusion coefficient, $\tilde{D}$, and the intrinsic diffusion
coefficient, D defined by

$$J = -D \ \frac{\partial C}{\partial x} , \tag{3.64}$$

are synonymous. In eqn 3.64 J is the flux of mobile atoms and C
is the concentration of mobile atoms per unit volume.

·In the case of compounds which are predominantly electronic
conductors or have metallic-like conductivity, a truncated form
of the Darken equation provides a relation between D^*, the tracer
diffusion coefficient and the thermodynamic activity, a, (O'Keeffe
1970 and Steele 1973)

$$\tilde{D} = D^* \ \left(\frac{\partial \ln a}{\partial \ln c} \right) , \tag{3.65}$$

and c is the mole fraction and is related to C by $C = \mathcal{N}c$ and
$\mathcal{N}$ is the number of all species per unit volume.

The activity a of a component can vary over many orders of magnitude in compounds with variable composition. In the vicinity of the stoichiometric composition the activity varies particularly strongly with composition and the thermodynamic driving force may be very large indeed, thereby leading to the prediction that $\tilde{D}$ will be very much larger than D^*. An example of this is provided by UO_{2+x} see fig 3.18.

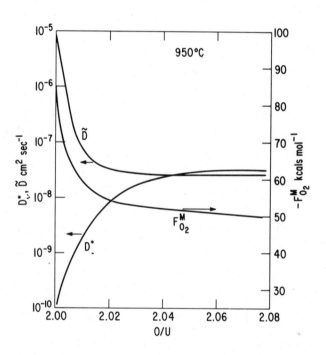

<u>Fig. 3.18.</u>   The partial molar free energy, $F^M_{O_2}$, the chemical diffusion coefficient, $\tilde{D}$ and the tracer oxygen diffusion coefficient D in UO_{2+x} (after Steele (1972)).

It has long been known that, in a rigorous derivation of the
Darken equation in *binary alloy systems* using the Onsager
phenomenological equations, an extra factor arises because of
non-zero Onsager cross coefficients (Howard and Lidiard 1964).
For the random concentrated alloy Manning (1968) showed that
the 'vacancy-wind-effect' was responsible for these non-zero
cross coefficients. Recently, Murch and Thorn (1979d) applied
the Onsager phenomenological equations appropriate for transport
in those highly defective solids which are predominantly electronic
conductors. They wrote for the fluxes of A atoms, tracer A atoms
(A*) and vacancies (V) respectively

$$J_A = L_{AA}X_A + L_{AA*}X_{A*} + L_{AV}X_V \qquad (3.66a)$$

$$J_{A*} = L_{A*A}X_A + L_{A*A*}X_{A*} + L_{A*V}X_V \qquad (3.66b)$$

$$J_V = -(J_A + J_{A*}), \qquad (3.66c)$$

where X_i = grad μ_i (μ_i is the chemical potential) and the L_{ij}
are the Onsager phenomenological coefficients. These equations
lead to the following expression for $\tilde{D}$

$$\tilde{D} = D_{A*} \left(\frac{\partial \ln a}{\partial \ln c_a} \right) \left[\frac{(L_{A*A} + L_{A*A*})c_A}{c_A L_{A*A*} - c_{A*}L_{A*A}} \right] \qquad (3.67)$$

where D_{A*} is the tracer diffusion coefficient for A^*
atoms and subscript a implies A + A*.
Eqn 3.67 is the rigorous Darken equation and differs from eqn
3.65 by the addition of the term in brackets [   ]. This finding
alone would hardly be of much consequence considering the

conclusions cited above in alloy systems but Murch and Thorn
(1979d) went on to show that the *same* correction term also occurs in t.
Nernst-Einstein equation

$$\frac{u_{A*}}{D_{A*}} = \frac{q_{A*}}{kT} \left[\frac{(L_{A*A} + L_{A*A}) \; c_A}{c_A L_{A*A*} - \dot{c}_{A*} L_{A*A}} \right],$$

(3.68)

where u_{A*} is the mobility of A* and A atoms and q_{A*} is the
charge on A* atoms (and A atoms). We have already seen in section
3.3.2 that the Path Probability Method of Sato and Kikuchi
(1971 a, b) and Monte Carlo (MC) computer simulations by Murch
and Thorn (1977a,d,e, 1978b) have shown that in the case of
single unit jumps (vacancy mechanism) the correction term [   ]
in eqn 3.68 can be identified with f_I/f where f is the tracer
correlation factor and f_I is the physical correlation factor.
For a detailed discussion of f_I see section 3.3.1. It suffices
here to remark that physical correlation effects, which are
a result of nonideality, affect both f and f_I but the ratio
is little affected. The Darken eqn 3.67 therefore becomes

$$\tilde{D} = D_{A*} \frac{f_I}{f} \left(\frac{\partial \ln a}{\partial \ln c_a} \right).$$

(3.69)

Murch (1980a) has subsequently verified eqn 3.69 by coupling MC
results of $\tilde{D}$ from a study of chemical diffusion using a steady-
state technique with previous computer calculations of the
quantities on the RHS of eqn 3.69.

We wish now to progress a little further from eqn 3.69.
We wish to write down a relation between the tracer diffusion
coefficient D_{A*} and the diffusion coefficient of *indistinguishable*
atoms, D_I. The latter diffusion coefficient is defined in a

way *identical* to $D_q^\dagger$, the diffusion coefficient of charge carriers
(eqn 3.58)

$$D_I = \tfrac{1}{2} \Gamma \lambda^2 f_I,$$ (3.70)

and from eqn 3.1 we find

$$\frac{D_{A*}}{D_I} = \frac{f}{f_I}.$$ (3.71)

Substituting eqn 3.71 into eqn 3.69 we find for the *rigorous* Darken
equation in the case of the vacancy mechanism (Murch and Thorn 1979d)

$$\tilde{D} = D_I \left(\frac{\partial \ln a}{\partial \ln c_a} \right).$$ (3.72)

The simplicity of eqn 3.72 is appealing; it reveals that a simple
and rigorous relation exists between the diffusion coefficient
in a compositional gradient, the diffusion coefficient in the
absence of the compositional gradient, and the thermodynamic
activity. It is of importance to note that D_I is not, itself,
directly measurable. The *only* diffusion coefficient which is
measurable in the absence of the *compositional* gradient is D_{A*}.

It is, of course, possible to gain access to D_I if the
material exhibits a measurable ionic conductivity and we refer
to section 3.3.1. From eqns 3.72 and 3.57 we can deduce the
following link between the Nernst-Einstein equation and the
Darken equation in the case of the vacancy mechanism and the
absence of neutral species which contribute to diffusion but not

---

†In fact, if the *atoms* are charge carriers $D_I = D_q$.

to ionic conductivity

$$\frac{u}{\tilde{D}} = \frac{q}{kT} \quad \left(\frac{\partial \ln a}{\partial \ln c_a}\right)^{-1}.$$ (3.73)

Thus in the case of nonstoichiometric compounds exhibiting predominantly electronic conductivity, the ionic mobility and the chemical diffusivity are directly related. Because of cancellation of correlation effects, it turns out that it is possible to arrive at eqn 3.73 without introducing correlation into the argument and this is the route that Wagner (1976) has taken. Through the use of eqn 3.73 one may determine a hypothetical u when $\tilde{D}$ is known e.g., in defective carbides and nitrides. Alternatively, in electronically conducting nonstoichiometric compounds, $\tilde{D}$ can be determined from a measurement of u if the electronic contribution to the conductivity can be suppressed.

Very recently, it has been possible to inspect the behaviour of $\tilde{D}$ in the case of nearest neighbour interactions, both repulsive and attractive (Murch 1980a). The results for the simple cubic lattice are given in fig. 3.19. For zero interactions it can be shown from eqns 3.69, 3.44 and 3.45 that $\tilde{D}$ is constant and this is verified by these computer calculations. For attractive interactions $\tilde{D}$ is depressed primarily because of the cooperative effect of the interactions as expressed through W. As the temperature is approached at which phase separation occurs then $\tilde{D} \to 0$ because $\frac{\partial \ln a}{\partial \ln c_a} \to 0$ and this is sketched *schematically* in the figure. This is a manifestation of the phenomenon of 'critical slowing down'. For repulsive interactions the behaviour is almost a mirror image of the attractive case. $\tilde{D}$ is enhanced, primarily because of the cooperative effect of the interactions as expressed through W. But at lower tempera-

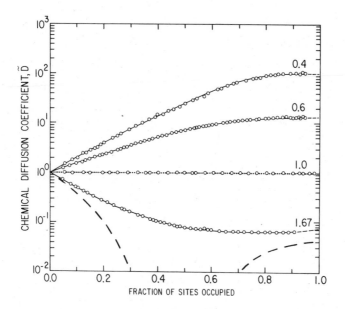

Fig. 3.19. The chemical diffusion coefficient, $\tilde{D}$, as a function of the fraction of occupied sites at various values of $\exp(-\varepsilon/kT)$ in the simple cubic lattice. Points are Monte Carlo estimates (Murch 1980a). The dotted line represents ideal mixing. The dashed line is the schematic behaviour *below* the critical temperature of unmixing.

tures the behaviour differs and also becomes more interesting. As we have already noted, for the simple lattices an ordered region is generated which is centered around $c_a = 0.5$. If we make an isothermal pass through that ordered region then the chemical potential behaves typically as shown in fig. 3.20 which is taken from a two dimension adsorption study by Murch (1980e). The rapid compositional change of the chemical potential in the ordered region leads to a very strong thermodynamic driving force which generates a bold maximum in $\tilde{D}$ within the ordered region, see fig. 3.21. If the ideal mixing thermodynamic driving force is substituted then we obtain the dashed line result which unequivocally reveals the important role of the chemical potential in the ordered region. These findings must now be put into a model framework for those electronically conducting 'nonstoichiometric' compounds such as the transition metal chalcogenides which give rise to homologous series of ordered phases at low temperature. Neglecting stress effects, section 3.4.3, one might expect a series of maxima in $\tilde{D}$ corresponding to the homologous series, but in the connecting two phase regions $\tilde{D} = 0$.

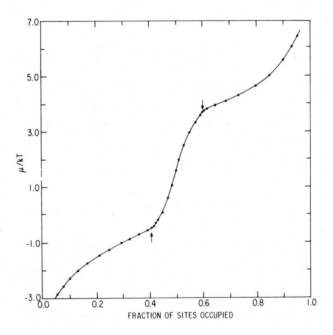

Fig. 3.20. Typical behaviour of the chemical potential for a cut through an ordered region. The data are for a square planar lattice with nearest neighbour repulsion and next nearest neighbour attraction between the particles (Murch 1980e). The arrows denote second order transitions.

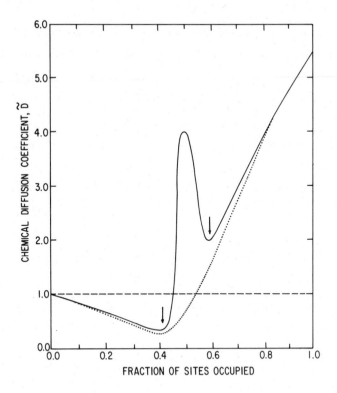

Fig. 3.21. Typical behaviour of the chemical diffusion coefficient for a cut through an ordered region. The solid curve uses the chemical potential of fig. 3.20 in the modified Darken equation 3.69, the dotted curve uses the ideal mixing chemical potential (eqn 2.5) (Murch 1980e).

So far we have discussed chemical diffusion in solids which are predominantly electronically conducting and where the atomic defects interact. We have also chosen to neglect the electronic flux because the *atomic* movement is rate determining. In actuality, in nonstoichiometric *ionic* compounds where the electrons and holes can be localized at sites the atomic flux and the flux of electronic defects also interact. Formally, this amounts to including in the phenomenological equations (3.66) the electronic flux and non-zero Onsager atom/electron cross coefficients but without leaving out J_V and J_{A*} (correlation effects) as Wagner (1976) and Yurek (1978) have done. Such a situation has not been modelled either . What is required is a time-dependent statistical mechanical analysis along the lines of the PP or MC methods which extends the equilibrium treatments recognizing coulombic interactions between atomic defects and electronic defects (Atlas 1968a, b, 1970). The equilibrium treatment of Atlas is, incidentally, suitable for calculating the thermodynamic factor in the Darken equation for such a case.

It is usual, however, to take a more naïve approach to the problem by assuming ideal mixing of the atomic defects and electronic defects. An example will illustrate the procedure. Consider the oxide $Co_{1-\delta}O$ with nonstoichiometry provided for by, for example, the reaction

$$Co_{Co} \rightleftharpoons V_{Co}^{\cdot} + h + Co(g). \qquad (3.74)$$

The thermodynamic term in the Darken equation (eqn 3.65) becomes, for an *ideal* mixture of charged vacancies and holes

$$\frac{\partial \ln a}{\partial \ln c} = \frac{2\,(1-\delta)}{\delta}. \qquad (3.75)$$

At this point one introduces a vacancy diffusion coefficient, D_V, which is related to $D*$ by

$$\frac{D*}{D_V} = \frac{\delta}{1-\delta}\ f. \qquad (3.76)$$

Eqn 3.76 depends on uncorrelated motion of vacancies, a vacancy

availability factor of δ and an *atom*[†]availability factor of
$(1-\delta)$. From eqns 3.75, 3.76 and 3.69 with $f_I = 1$ (for no interactions) we find

$$\overset{\approx}{D} = 2D_v .\qquad(3.77)$$

That is to say, in the absence of interactions, the chemical diffusion coefficient is directly related to the vacancy diffusion coefficient. The factor 2 is thus transmitted as a consequence of ideal mixing of equal concentrations of the two species: $V_{Co}^{\cdot}$ and holes. For other reactions and hence different relative concentrations of the two species the numerical factor may change.

One may derive eqn 3.77 in quite a different way by making use of Wagner's (1930) analysis of the simultaneous diffusion of ions and electrons. Since the hole is more mobile than the cobalt vacancy, it will tend to run ahead creating a diffusion potential, ϕ, formally a space charge or 'Nernst field' due to the incipient separation of charges which arises because of the differing mobilities. We write for the fluxes of the vacancies and holes

$$J_{V_{Co}^{\cdot}} = -D_v \frac{\partial C(V_{Co}^{\cdot})}{dx} + \left(\frac{zqD_v C(V_{Co}^{\cdot})}{kT}\right)\frac{d\phi}{dx}\qquad(3.78a)$$

and

$$J_h = -D_h \frac{dC(h)}{dx} - \left(\frac{zqD_h C(h)}{kT}\right)\frac{d\phi}{dx},\qquad(3.78b)$$

where $z\,(=1)$ is the number of charges on the defect (ion or

---

[†]The atom availability factor is analogous to the vacancy availability factor, see section 3.2.1.

electronic defect) and the D's are defined in the *absence* of a compositional gradient and D_h is defined in a way analogous to D_v. Electroneutrality requires that

$$J_{v^{\bullet}_{co}} = J_h \qquad (3.79)$$

and since $D_h \gg D_v$

then $\tilde{D} = 2D_v$ $\qquad\qquad (3.80)$

which is the same result as eqn 3.77. It is sometimes said that the factor 2 is a *result* of the diffusion potential. As we have seen above the factor 2 can also be thought of as a consequence of ideal mixing of the two species. Both points of view are correct.

In the more general case where the electronic defect mobility is comparable with the ion defect mobility the equation for the chemical diffusion coefficient for the reaction (3.74) will be of the form (Heyne 1968, Steele 1972)

$$\tilde{D} = \frac{2D_v D_e}{D_v + D_e} \qquad (3.81)$$

where D_e is the diffusion coefficient of the electronic defect. It will also be apparent that in the case where *ionic* conductivity predominates then the equation for the chemical diffusion coefficient reads

$$\tilde{D} = 2D_e \qquad (3.82)$$

3.4.3 Chemical Diffusion and Stress

O'Keeffe (1970) has remarked that when the composition of
a nonstoichiometric compound like $Fe_{1-\delta}O$ is suddenly changed
near the surface, a stress will develop since there will inevitably
be a change in lattice parameter. The stress could then result
in a component of diffusion normal to the concentration gradient
to relax the stress and possibly to change the shape of the
specimen. O'Keeffe chose to include the effect of stress in
the chemical potential. He modified the Darken equation (eqn
3.65 with the *ad hoc* addition of $\frac{1}{f}$) to

$$\tilde{D} = \frac{D^*}{f} \left[\frac{d\ln a}{d\ln c} + \frac{1}{RT} \frac{d\delta\mu}{d\ln c} \right], \tag{3.83}$$

where $\delta\mu$ is the change in the chemical potential due to the
stress. O'Keeffe (1970) estimated the second term in brackets
in eqn 3.83

$$\frac{1}{RT} \frac{d(\delta\mu)}{d\ln c} = \frac{\bar{V}}{\beta RT} \frac{d(\Delta V/V)}{d\ln c}, \tag{3.84}$$

where $\Delta V/V$ is the fractional change in volume that would occur
in an unstressed crystal upon changing concentration, β is the
compressibility and $\bar{V}$ is the molar volume. An order of magnitude
calculation revealed that eqn 3.84 equals about 20 in $Fe_{1-\delta}O$
which is comparable to the first term in brackets in eqn 3.83
which is about 30. The value ascribed to eqn 3.84 seems to be

much too high since *without* the stress term, $\tilde{D}$ as calculated for
$Fe_{1-\delta}O$ is already about 9-50 times higher than the measured $\tilde{D}$
(O'Keeffe 1970). More recently, Yurek (1978) has indicated,
without specifying his sources of experimental data, that the
calculated $\tilde{D}$ can be brought into coincidence with $\tilde{D}$ as measured
with f = 0.721. Strictly, it is f/f_I = 0.721 but that is another
matter, see section 3.4.2. The point here is that a stress term
does not need to be invoked for the Darken equation to hold in
$Fe_{1-\delta}O$.

O'Keeffe notes that the first term in [   ] in eqn 3.83 will
be much larger than the stress term for nonstoichiometric
compounds close to stoichiometry. For systems where the component
partial pressure isotherms are rather flat e.g., interstitial
solid solutions just above the critical temperature, the
thermodynamic term will approach zero[†] and the stress term may
predominate. Activity gradients calculated from equilibrium
activities may not then provide an accurate measure of the driving
force for diffusion in a compositional gradient.

3.4.4 Chemical Diffusion and Crystallographic Shear

The early work in many nonstoichiometric transition metal oxides
often revealed apparently extensive ranges of compositional homogenei
The emphasis is on the word 'apparently' because many of these wide
composition fields have in recent years been reduced to a series of
ordered shear plane phases. A quantitative and general theory

---

[†]This is a manifestation of the phenomenon of critical slowing down.

of diffusion and chemical diffusion in particular in such systems
has not yet been developed. Most of our understanding is
conjecture based on observations through electron microscopy.
Our discussion here is meant as an introduction to the field
and follows expositions of O'Keeffe (1970, 1971).

Magnéli (1953) and coworkers originally showed that in many higher
oxides of the transition metals there exists a homologous series
of oxides related by a simple structural principle to some parent
oxide. Famous examples are the series M_nO_{3n-1} or M_nO_{3n-2}
derivable from the ReO_3 structure and M_nO_{2n-1} from rutile.

In each of the phases the structure may be considered as
consisting of slabs of the parent structure which are separated
by planes containing a higher cation concentration. The geometry
of the situation may be appreciated by conceiving of slices in
the crystal at regular intervals along a certain plane. We then
imagine that some of the anions are removed and that the crystal
is sewn up by shear of the two pieces along the cut. Fig. 3.22
illustrates such a shearing process. The plane of shear is referred
to as a Crystallographic Shear (CS) plane. It can now be seen
that the crystal structure is determined by regular spacings of
CS planes with the *composition* dependent on the nature and
spacing of the CS planes. Much structural development has taken
place in this field and we refer to reviews by Wadsley (1963),
Tilley (1972), Anderson (1972), Anderson and Tilley (1974) and a
conference (Eyring and O'Keeffe 1970).

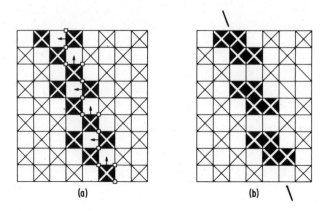

(a)   (b)

Fig. 3.22.   A sketch of the formation of intermediate phases by
crystallographic shear, one layer of MO_6 octahedra showing *corner*
sharing of oxygen atoms. Oxygen vacancies shown by (O) tend to
distribute themselves in the configuration shown. The black octahedra
can then move to the positions shown in (b) in the which the vacancies
have been annihilated by *edge*-sharing between adjacent octahedra
(after Wadsley 1963).

An important development in the field was the contribution
of Wadsley (1963). He proposed that in a nonequilibrium situation
individual CS planes may well lead quite an independent existence
and, further, that apparently slight nonstoichiometry of a phase
could be traced to a spacing of CS planes which is not entirely
regular. Such CS planes are now called "Wadsley defects' and have
been confirmed by the electron microscope technique of lattice
imaging which takes advantage of the region of different density
at the shear plane.

Anderson and Hyde (1967) proposed that changes in composition
are accomodated by isolated Wadsley defects which can augment
or diminish by way of diffusion in and out of the crystal by
anion vacancies. Accordingly, the Wadsley defect acts as a source

or sink for point defects.   In this case one might expect a correlation between chemical diffusion and oxygen tracer diffusion. Electron microscope investigations on NbO_2F and rutile seem to verify the mechanism.   Anderson and Tilley (1972) have pointed out that shear planes tend to grow in from a free surface and O'Keeffe (1971) has suggested that the disordered region bounding the surface of crystallographic shear would then provide a favourable diffusion path for oxygen ions leaving the crystal.   As a consequence, as long as a shear plane intersects the surface of the crystal, no anion mobility in the parent crystal is required, in contrast to the original model proposed by Anderson and Hyde.

In an alternative explanation for changes in composition, Andersson and Wadsley (1966) suggested that the Wadsley defects themselves form at the surface of the crystal and migrate into the interior by means of cooperative jumps of both metal and oxygen ions in equal amounts, Fig. 3.23.   This mechanism allows a change of composition without the direct participation of point defects. This is probably the first example of a change in chemical composition in the 'absence' of diffusion.   Because of the unlikelihood of acquiring sufficient thermal energy, Wadsley defects can be expected to be immobile in the absence of a driving force.   A strong driving force is, however, provided by the interplanar interaction which ultimately results in ordering into a homologous series.   Bursill and Hyde (1971) have deduced the form of the interaction potential between (132) CS planes in reduced TiO_2. They revealed that the interaction potential was repulsive at close approach, exhibited a shallow minimum at about 7.5 nm and a weak attraction at greater distances.   At equilibrium, closely spaced CS planes will therefore tend to have a uniform spacing.

122

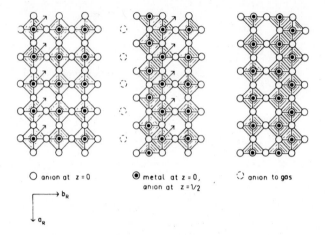

O anion at z = 0

⦿ metal at z = 0,
anion at z = 1/2

◌ anion to gas

→ b_R

↓ a_R

<u>Fig. 3.23</u>. The Andersson/Wadsley mechanism of CS plane migration
showing the production of an 010 CS plane at the surface of a
crystal of ReO_3. Firstly, (a,b), anions escape to the gas phase,
then rows of cations and anions cooperatively 'diffuse' inwards
from normal sites to interstices as indicated by the arrows.
(c) Next, subsequent lateral migration of the CS plane is by
cooperative 'diffusion' of MO rows. In its wake ReO_3 is
regenerated but displaced by a/2.

It is clear that with the Andersson and Wadsley (1966)
mechanism there could be a large chemical 'diffusion' coefficient as
atoms are swept through the crystal by a flux of Wadsley defects
but a small tracer diffusion coefficient since at equilibrium the
CS planes are immobile and atoms can then migrate only by
the intrinsic point defects which are probably in low concentration.

Merritt (1970) and Bursill and Hyde (1971,1972) have shown that
the attainment of equilibrium in TiO_x 1.5<x<2.0 is generally
extremely slow and that there is much hysteresis. This has been

interpreted as being due to a kinetically hindered creation and/or migration of CS planes. However, over a small range of composition $TiO_{1.90}$ - $TiO_{1.937}$ change of composition is provided by a very rapid change, not in the number, but in the *orientation* of the CS planes. The orientation process is accomodated by a relatively small number of atomic jumps. No hysteresis was observed within that composition range. Accordingly, we might expect the chemical diffusion coefficient to vary quite considerably over the composition range $1.5 < x < 2.0$ in TiO_x.

We remarked at the beginning of this section that no theoretical treatment has yet been advanced to explain and quantify the above observations. We suggest that some significant progress at least for the mechanism of Anderson and Hyde (1967) can probably be made along the lines of computer simulation of dislocation movement. See, for example, a conference on computer simulation (Arsenault et al., 1976).

3.5   Two Recent Theoretical Methods

3.5.1   General Remarks

Within about the last decade two powerful methods have emerged for use in dealing with kinetic processes in cooperative systems. The first of these is the Path Probability method developed by Sato and Kikuchi. The method is a time-dependent version of the well-known cluster variation method. The second method is the Monte Carlo method which has been extended into the time-dependent regime principally by the author. The two methods enjoy a particularly happy partnership since the numerical method provides an 'exact' reference point for the development of the ultimately preferable analytical approach. In the following two sections we describe these methods in turn.

3.5.2   The Path Probability Method

The Path Probability (PP) method is a very powerful analytical technique for dealing with the *kinetics* of cooperative systems. (Sato and Kikuchi (1969, 1970, 1971a, b, 1976, 1977, 1979)). The method is by far the most comprehensive and powerful analytical tool available for dealing with atomic transport in highly defective solids. In the author's opinion it is likely that the method will continue to make a major impact on our understanding of the subject.

Firstly, it is important to recognize where the PP method stands in relation to previous methods in diffusion theory such as the well known "pair-association method" of Lidiard (1955). In highly defective systems and concentrated alloys as well, the

basic premise is that mutual interactions among the atoms affect
the diffusion process. The problem can be restated as a time-
dependent many-body problem which is amenable only to a statistical
mechanical analysis. The pair-association method of Lidiard (1955),
as conventionally stated, does not contain any statistical mechanics
and cannot describe such problems. On the other hand, the PP method
was formulated to deal specifically with these problems. It may be
pictured as a time-dependent statistical mechanical method which
has been formulated by adding a space axis to the cluster variation
method of equilibrium statistical mechanics, Kikuchi (1951,1966).
The PP method certainly makes use of some of the strategy of
Lidiard's pair-association method and to this extent may be viewed
as a generalization of that method.

The PP method was developed as a *time-dependent* version of
the Cluster Variation (CV) method (Kikuchi 1951) in order to
deal with problems in irreversible statistical mechanics. The
CV method itself is a hierarchy of approximations *in closed form*.
In this method, the variational variables are based on certain
types of basic atom clusters and a better approximation in the
hierarchy is generally associated with making use of larger
clusters. A characteristic of the CV method is the formulation of
the configurational entropy of the system from the number of
possible rearrangements of the clusters but without creating
'inconsistencies'. The smallest cluster is a single lattice point
and this 'point' approximation is equivalent to the Bragg-Williams
approximation, see section 2.2. The next larger cluster is
a pair of lattice points and this 'pair' approximation is quite
equivalent to the Bethe approximation, see also section 2.3.
Larger clusters such as the tetrahedron have also been used but

unless one is interested in very accurate critical properties, such clusters are really only necessary in the f.c.c. structure. In that case, because two nearest neighbours of a point can themselves be nearest neighbours the pair approximation is quite poor. The same objection can also be raised concerning the triangular lattice. For other lattices, the pair approximation is adequate for most practical purposes.

As we noted above, the PP method was developed to deal specifically with cooperative phenomena in nonequilibrium systems. The problem of the description of tracer diffusion and ionic conductivity in a lattice gas, i.e., a gas of interacting particles (ions or defects) distributed over a rigid array of potential wells, is an example of a suitable application of the method. As an aside, we should note that for nearest-neighbour interactions the problem is, in fact, a restatement of a time-dependent Ising model where the magnetization is conserved.

In general terms, the PP method is invoked in the following way. The equilibrium distribution of ions, and this includes tracers and vacancies, is described firstly in terms of the CV method. The deviation from the equilibrium distribution of ions by the presence of a chemical potential gradient is then determined. Ions flow along the gradient of chemical potential but in such a way that the overall distribution of ions is unchanged. The deviation from equilibrium and the rate of flow of the ions are then determined with the PP method.

The general mathematical strategy of the PP method is as follows. A function, analogous to the partition function of equilibrium statistical mechanics is sought whose extremum value corresponds to the manner in which the system changes according to a diffusive process. In essence, a set of variables $\{A_i\}$ which specifies the change of state of the system in Δt is defined and a path probability function, $P(A_1, A_2 ...)$ that the change of $\{A_i\}$ occurs is then determined. The values of A_i which lead to a maximum in P determine the direction in which the system changes.

We now wish to describe the application of the PP method to a particular lattice gas. The approximation level is the 'pair' approximation. We consider a lattice of N *equivalent* sites *a priori* with a coordination Z and containing ions which interact only as nearest neighbours. We *define* an interaction energy, ε, ($\varepsilon < o$: repulsion and $\varepsilon > o$: attraction).

To develop the application of the PP method we shall introduce the concepts of state and path variables. Now these quantities are not unique to the PP method. Indeed their analogues can be found in the pair-association method. We have chosen, however, to retain the definitions as given by Sato and Kikuchi.

State Variables

The fraction of lattice points occupied by host ions, A, is defined by p_A, of tracer ions, A*, by p_{A*} and of vacancies by p_V; p_A, p_{A*} and p_V are the *state* variables associated with the *point*. For a *pair* of points the fraction of the possible configurations are y_{AV}, y_{A*V}, y_{VV}, etc. These are termed the *state* variables associated with a *pair* of points. One can consider larger clusters but the mathematics becomes quite tedious. In the *pair* approximation, state variables for a pair and site alone are used. The set of state variables $\{y\}$ specifies a given state of the system.

Once the set $\{y\}$ is defined, we can formulate the probability $\chi\{y\}$ that the system takes a state $\{y\}$ when the system is in isolation or in contact with a heat bath. The equilibrium state of the system, $\{y_e\}$ is the state which maximizes $\chi \{y\}$. This is, of course, the usual conceptual procedure of equilibrium statistical mechanics.

We now generalize the state variables by introducing the concept of *path* variables.

Path Variables

We define $Y_{A*V,VA*}^{(n)}$ (t, t + Δt) as the probability of finding a situation at n, a bond connecting the m'th and m + 1 th points, such that the configuration is A*-V (tracer A-vacancy) at time t and V-A* at time t + Δt. Similarly, we may define a probability for the reverse jump by $Y_{VA*,A*V}^{(n)}$ (t, t + Δt). It is immediately clear that the difference between these two probabilities is the net probability or net flux that a tracer A* moves from point m to m+1 in time Δt

$$J_{A*}^{(n)} = Y_{A*V,VA*}^{(n)}(t, t + \Delta t) - Y_{VA*,A*V}^{(n)} (t, t + \Delta t). \qquad (3.85)$$

At equilibrium it is clear that $J_{A*}^{(n)} \equiv 0$.
Let us now examine tracer diffusion.

Path Variables and Tracer Diffusion

For a small tracer gradient $dp_{A*}^{(m)}/dm$, $J_{A*}^{(n)}$ does not vanish but is proportional to that gradient. The constant of proportionality is the tracer diffusion coefficient, D_{A*}, of A*, apart from a proportionality constant, c

$$cJ_{A*}^{(n)} = - D_{A*}dp_{A*}^{(m)}/dm \qquad (3.86)$$

Our goal then, is to find the relation between $J_{A*}^{(n)}$ and $dp_{A*}^{(m)}/dm$.

So far, we have said nothing fundamentally new that was not known before the appearance of the PP method. We have merely set the stage for the solution of the problem by the PP method. For this solution the PP method asks only two questions. What is the

most probable path? What are the state variables at steady state?
Let us answer the first question.

The Path Probability

We wish to gain access to the probability that the system changes
from a state $\{y(t)\}$ at time t to some other state $\{y(t + \Delta t)\}$ in a time
Δt by a diffusive process. We write this probability as $P(\{y(t)\},$
$\{y(t + \Delta t)\})$. It is, of course, kinetically possible to change into
any one of a large number of different states. But we are interested
in that state to which the system is *most likely* to change. This
problem cannot be answered within the framework of equilibrium statistical mechanics. It may, however, be answered using probabilistic
reasoning provided a diffusive process is given.

Rather than write P in terms of the state variables, y, it
turns out to be considerably more convenient to write P in terms
of the path variables, Y. Now, in principle, in the implementation
of the PP method, one would maximize P with respect to Y but keep
the y's constant. The maximization process then would lead to a
relation between $\{Y(t,t+\Delta t)\}$ and $\{y(t)\}$. This is a lengthy process
however, but there exists an important short cut which greatly
simplifies the entire application of the PP method. The short cut
makes use of a superposition principle (Kirkwood 1935) which says,
in essence, that by assuming that the exchange frequencies are
dictated by the environment *before* the atom leaves, and that pairs
of sites, the y's, are assumed independent (the pair approximation)
then the Y's can be written down by inspection as

$$Y_{A*V,VA*}^{(n)}(t,\ t + \Delta t) = \Delta t \nu \exp(-E^\circ/kT) W_{A*_V,VA*}^{(n)}(t) y_{A*V}^{(n)}(t) \qquad (3.87a)$$

and

$$Y_{VA*,A*V}^{(n)}(t, t + \Delta t) = \Delta t \nu \exp(-E^\circ/kT) W_{VA*,A*V}^{(n)}(t) y_{VA*}^{(n)}(t), \qquad (3.87b)$$

where ν is the vibrational frequency, E_{act}° is the activation energy for diffusion of an isolated ion. The W's are the effective frequency factors which take into account the breaking of bonds when a tracer ion jumps, i.e., $W_{A*V,VA*}^{(n)}$ depends on $y^{(n-1)}$ and $y^{(n)}$ while $W_{VA*,A*V}^{(n)}$ depends on $y^{(n+1)}$ and $y^{(n)}$. In other words, W is given by the state immediately *before* the ion jumps.

It is usually implied that eqns 3.87a,b are valid only for the pair approximation (Sato and Kikuchi 1971a,b). It seems to this author that these equations are probably more generally valid to the extent that the only assumptions are 1) that the environment of the atom determines the frequency and 2) that there is pair-wise additivity of interaction energies. But this remains to be proved.

Now, eqns 3.87a,b are valid for *any* given state at time t. Indeed we should note that these equations could have been written down within the framework of the pair-association method. The primary theoretical task in the PP method is to find a relation between the tracer flux, J_{A*} and the chemical potential gradient $d\mu/dm$. The form of this relation, which is unique to the PP method, is too unwieldy to give here. It is expressible in terms of the local equilibrium values of the y's (in the absence of the gradient) and their values under the influence of the gradient. To use this relation one final condition of constraint is provided by the condition of steady state.

The Steady State

There are, of course, many ways of representing steady state.
Sato and Kikuchi choose to use

$$y_{A*V}^{(n)} (t + \Delta t) - y_{A*V}^{(n)} (t) = 0 . \qquad (3.88)$$

which simply states that the nearest neighbour relation does not
change with time. Both $y_{A*V}^{(n)}(t+\Delta t)$ and $y_{A*V}^{(n)}(t)$ can be written in
terms of the appropriate path variables $\{Y(t,t+\Delta t)\}$ which themselves
are expressible in terms of the state variables $\{y(t)\}$ at time t.
In effect, as we mentioned above, eqn 3.88 determines the most
probable set of stable variables in the steady state. This permits
the construction of the standard linear form of the flux equation,
see eqn 3.66b and this leads, of course, directly to an expression
for D_{A*}. It takes the form:

$$D_{A*} = \frac{\lambda^2}{2} \nu \exp(-E°_{act}/kT) VWf \qquad (3.89)$$

where f is the tracer correlation factor which is a function of the
concentration of diffusing ions and ε/kT, see section 3.2.3. For
diffusion theory, the new factors in eqn 3.89 are V and W. V is
termed the "vacancy availability factor" and is given by

$$V = y_{A*V,e}/P_{A*,e}, \qquad (3.90)$$

V is an "effective" vacancy concentration factor. It describes the
accessibility of the ions to the vacancies in the interacting
system. W is called an effective frequency factor and is given by

$$W = [(y_{AA*,e} + y_{A*A*,e}) \exp(-\varepsilon/kT) + y_{VA*,e}]^2 P_{A*,e}^{-2} \qquad (3.91)$$

W reflects the enhancement or depression of the exchange frequency, $\nu \exp(-E^\circ_{act}/kT)$, caused by the binding (positive or negative) of the neighbours to the diffusing ion. In eqns 3.90 and 3.91 the subscript e signifies an equilibrium value.

Ionic Conductivity

In formulating ionic conductivity an external electric field, F is imposed on the system and the flow of ions in the field direction is calculated. In this case, analogous to eqn 3.85, we have that

$$J_A^{(n)} = Y_{AV,VA}^{(n)} (t, t + \Delta t) - Y_{VA,AV}^{(n)} (t, t + \Delta t) \tag{3.92}$$

The ionic conductivity σ is the ratio of $J_A^{(n)}$ to the applied field, F

$$c' J_A^{(n)} = \sigma F \tag{3.93}$$

and c' is the normalization constant.

Our goal here is to find the relation between $J_A^{(n)}$ and F. This is quite analogous to our case of tracer diffusion cited above. As with D_{A*} the first of the two problems concerns the expression of the Y's in terms of the most probable path; the second is to find the state variables in the stationary state. The treatment is parallel to that already given and we simply quote the final expression for σ

$$\sigma = \frac{\eta e^2}{kT} \frac{\lambda^2}{2} \nu \exp(-E^\circ_{act}/kT) \ VWf_I \tag{3.94}$$

where η is the number of ions per unit volume and f_I is the physical correlation factor (section 3.3.2).

Finally, we point out that the chemical diffusion coefficient, $\tilde{D}$, is also immediately accessible since (Murch and Thorn 1979d) (see section 3.4.2)

$$\tilde{D} = D_{A*} \frac{f_I}{f} \left(\frac{\partial \ln a}{\partial \ln p_A} \right) \tag{3.95}$$

and a is the activity and is easily obtainable by making use of the
equilibrium pair approximation of the CV method.

3.5.3   The Monte Carlo Method

The label: "Monte Carlo" (MC) method is commonly used in science
to denote the use of computer generated pseudo-random numbers for
the purpose of estimating multi-dimensional integrals. Often in
the process of integration a physical process may be simulated,
but this is by no means invariably the case.

In this section we have restricted ourselves to a review of
the MC method as developed for the thermodynamics and diffusion in
highly defective solids (Murch and Thorn 1977a, b, c, d, e,
1978b, 1979, a, b, c and references therein). For reviews of the
MC method in other areas of statistical physics we refer to Valleau
and Torrie (1976), Valleau and Whittington (1976), Wood and Erpenbeck
(1976), and Binder (1979).

At present, the MC method provides a powerful alternative to
the Path Probability method (see section 3.5.2). It is expected
that the MC method will continue to be exploited in order to provide
an exact reference base and, indeed, an incentive for the analytical
development of the subject.

For our purposes we may identify two distinct forms of application of the MC method. In the first, we are interested in obtaining
equilibrium thermodynamic averages of observable mechanical quantities.
In the second, we are interested in obtaining estimates of non-
equilibrium quantities. Let us now focus on the first of these,
initially in general terms, but later in more detail.

Equilibrium Properties

We wish to obtain the thermodynamic average of some observable property, A, which is a function of the configuration of the particles (atoms or defects) in the system. In order that the problem can be handled by the MC method we can consider only a finite number of particles, usually somewhere between 10^3 and 10^5; this is inevitably a compromise between one's computing budget and the precision hoped for. In order to eliminate edge effects the computer 'box' which contains the particles is made periodic. That is to say, the box is surrounded by periodic images of itself.

There are two basic types of boxes for MC calculations. The first allows for each of the particles to occupy any position within the box: this is a fluid-type MC calculation. The second allows for the particles to occupy only a regular array of sites: this is an Ising or lattice gas-type MC calculation. Calculations in highly defective solids have so far been restricted to lattice gases.

If the positions of, say, N particles are known in the box then we can easily calculate the potential energy, E, of the system in state k

$$E_k = \frac{1}{2} \sum_{i=1}^{N} \sum_{j=1}^{N} \phi_{ij} \qquad i \neq j \qquad (3.96)$$

where ϕ_{ij} is the pair potential between particles i and j. Eqn 3.96 assumes, of course, pairwise additivity.

To calculate the properties of the system we often use, for convenience, the petit canonical (N,V,T) ensemble. The equilibrium value of the quantity of interest, A, is given, for the lattice gas, by

$$<A> = \sum_i P_i A_i \tag{3.97}$$

where the summation is over all possible configurations and the probability density P_i is given by

$$P_i = \frac{\exp(-E_i/kT)}{Q} \tag{3.98}$$

where Q is the petit canonical configurational partition function given by

$$Q = \sum_i \exp(-E_i/kT) . \tag{3.99}$$

The obvious method of calculating eqn 3.97 is to put each of the N particles into the box randomly, then to calculate A_i and also the energy of this configuration (the latter according to eqn 3.96) and assign this configuration a weight $\exp(-E_i/kT)$. This process would then be repeated. This method is unsuccessful for the following reason. With a high probability a configuration will be generated where $\exp(-E/kT)$ is very small and hence the configuration has very low weight. Consider, for example, the likelihood of generating a configuration with no overlaps for a hard sphere gas at liquid densities. Thus a very unreliable estimate of <A> would result. In actual fact, only a quite small part of configuration space is actually *effective* in evaluating thermodynamic averages and we can make use of this fact in the following elegant method.

The usual method used for calculating eqn 3.97 is one introduced by Metropolis et al. (1953). What one does is to concentrate the sampling within the region of configuration space which is effective in the average. This is known as 'importance sampling'. That is to say, instead of choosing configurations randomly and then weighting them with exp(-E/kT), we choose configurations with a probability of exp(-E/kT) and weight them evenly.

Successive sample configurations are not in fact chosen independently but rather form an irreducible Markov chain in which every configuration is a consequence of another configuration. The N particles are placed in the box in an arbitrary configuration. The particles are chosen in succession or randomly and then hypothetically moved according to the following prescription

$$X \to X \pm \alpha r_1 \tag{3.100a}$$

$$Y \to Y \pm \alpha r_2 \tag{3.100b}$$

$$Z \to Z \pm \alpha r_3 \tag{3.100c}$$

where α is the maximum allowed displacement and the r's are pseudo-random numbers uniform on the interval (0,1]. After a particle is moved it can be found equally likely anywhere in a cube of length 2α. One notes that because of the periodic boundary, a move 'outside' the box merely means that the particle reenters through the opposite face of the box. Now, the *change* in energy, ΔE, of the system as a result of the move is calculated. If $\Delta E < 0$, i.e., if the move would lower the energy of the system, the move is permitted. If $\Delta E > 0$ then the move is allowed with probability exp(-ΔE/kT). A pseudo-random number, r_4, is generated, uniform on the interval (0,1]. If $r_4 < \exp(-\Delta E/kT)$ the move is permitted. If $r_4 > \exp(-\Delta E/kT)$ the particle is returned to its original position. Whether or not the move is successful the resulting configuration is considered a new one for the purposes of taking averages. After M such moves the thermodynamic average of the quantity A is calculated from

$$<A> = \frac{1}{M} \sum_{j=1}^{M} A_j \; .$$

(3.101)

The transition probabilities to move from configuration to configuration given above are not unique but even after almost 30 years, they have proved to be the optimum choice in nearly all implementations of the MC method. That these transition probabilities lead to a canonical distribution can also be readily proved and we refer to the original paper by Metropolis et al. (1953) and also more recent and sophisticated analyses by Binder (1979) and Valleau and Whittington (1976). The value of α in eqns 3.100a,b,c, although arbitrary, is usually chosen such that about half the moves are accepted. This ensures a fairly rapid exploration of configuration space. Typically about 10^5-10^6 moves are required for an adequate sampling of configuration space but this number may not be nearly enough in 'condensed' systems.

The estimation of errors is quite a difficult problem because of the obvious strong correlation between successive configurations. The usual procedure is to form a sequence of estimates of <A> over blocks of configurations once the initial transient from an arbitrary starting configuration (fig. 3.24) has disappeared. The block size

is chosen large enough such that there is no correlation between the
block estimates of <A>.

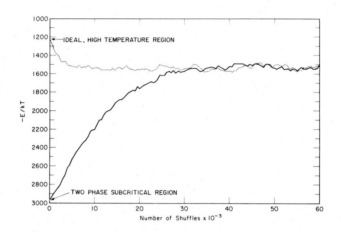

<u>Fig. 3.24.</u>   The approach to thermal equilibrium from both a random
distribution and a two-phase distribution.

In highly defective solids one is particularly interested in the
structure sensitive quantities such as the heat capacity and the
chemical potential.   The former can be found easily from the energy
fluctuations

$$C_v = \frac{<E^2> - <E>^2}{kT^2} \qquad\qquad (3.102)$$

bearing in mind our comments above on the estimation of errors.
The chemical potential in the lattice gas can be readily obtained
by either of two methods.   In the first, we write μ as

$$\mu = (\frac{\partial F}{\partial N})_{B,T} \qquad\qquad (3.103)$$

where B is the number of lattice sites and F is the configurational
free energy and noting that F is given by

$$F = -kT\ln Q \qquad (3.104)$$

then the derivative of eqn 3.103 can be replaced by its finite
difference form

$$\mu = \lim_{N \to \infty} -kT\ln\left[Q(N+1,B,T)/Q(N,B,T)\right] \qquad (3.105)$$

and μ can then be expressed as (Murch and Thorn 1978d)

$$\mu = -kT \ln \frac{\sum_{\ell=1}^{M}\left[\sum_{i=1}^{B}\exp(-\zeta_i/kT)\right]_\ell}{(N+1)M} \qquad (3.106)$$

where ζ is the hypothetical energy of interaction when a hypothetical
particle, the (N+1)th, is added to a frozen configuration ℓ characteristic of equilibrium (Widom 1963).

Alternatively, the density, N/B, consistent with an applied
chemical potential (formally equivalent to the magnetization and
external magnetic field of the Ising model respectively), can be
obtained by recasting the system as a grand canonical (μ,B,T)
ensemble (Chestnut and Salsburg 1963, Murch and Thorn 1978c). In
this case, the thermodynamic average of some observable, A, is
given by

$$\langle A \rangle = \sum_{i,N} P_i(N)A_i(N) \qquad (3.107)$$

and $P_i(N)$ is the probability that a configuration chosen at random
from the grand ensemble will contain N particles and have energy
E_i. P_i is given by

$$P_i = \frac{\exp(-E_i/kT)\exp(N\mu/kT)}{\Xi} \qquad (3.108)$$

where Ξ is the grand canonical configurational partition function

$$\Xi = \sum_N \exp(N\mu/kT)Q \qquad (3.109)$$

and Q is given by eqn 3.99. In this case the weight assigned to each configuration is given by

$$u_i = \exp(-(E_i - \mu N_i)/kT). \qquad (3.110)$$

A starting configuration is obtained by placing an *arbitrary* number of particles randomly in the lattice. Each additional configuration is generated in the following way. A site is selected at random and a *possible* occupant i.e., particle or vacant site is also selected. If the selected site happens to have the same 'occupant' as that generated, the next configuration is, of course, identical with the original configuration. Otherwise, the quantity $[(E_2 - \mu N_2) - (E_1 - \mu N_i)]/kT$ is computed where 2 refers to the new configuration and 1 to the old. If $(E_2 - \mu N_2) < (E_1 - \mu N_1)$ then the new configuration is accepted (this involves either removal or addition of a particle). If $(E_2 - \mu N_2) > (E_1 - \mu N_1)$ then the ratio u_2/u_1 is calculated and compared with a random number, r, uniform on the interval $(0,1]$. The new configuration is accepted only if $r < u_2/u_1$.

A particular advantage with this ensemble is the exponentially rapid approach to equilibrium in contrast with the petit canonical ensemble. In fact, for diffusion simulations in ordered structures, it is preferable to use this ensemble to generate a configuration characteristic of equilibrium and then to switch to the petit canonic ensemble for the diffusion part of the calculation where it is necessary to have a constant number of particles.

So far we have not commented on the form of the pair potential. For short range potentials, the application of the MC method is straightforward. The handling of ionic systems with say, *charged* point defects or defect clusters and their valence defect complements is more difficult because the MC system is necessarily finite. It is important to recognize that the existence of periodic boundaries does *not* imply that the system is infinite. Some of the various methods used to cope with Coulombic potentials are detailed by Valleau and Whittington (1976).

Interstitial defects may be treated as occupying sites of higher potential energy a priori. A calculation of this type has been performed in β-alumina (Murch and Thorn 1977d). Systems with a high concentration of defects induced by a high level of dopant e.g., CeO_2 doped with yttria, calcia stabilized zirconia etc. may be handled by initially distributing the lower valent cations at random and then using this frozen configuration for the low temperature equilibrium distribution of the anion vacancies. A calculation of this type has been performed by Murch and Nowick (1980). Alternatively, equilibrium at high temperature may be achieved by using both defect components. A configuration characteristic of this equilibrium may then by quenched to a lower temperature and the anion vacancies re-equilibrated to the frozen distribution of lower valent cations. In both cases, at low temperatures the immobile lower valent cations act as pinning points for the highly mobile anion vacancies.

These comments are only meant as suggestions for situations where the author believes that the MC method is likely to prove useful and *feasible* on present generation computers. Future developments in computer technology can only lead to a more adventurous choice of simulation conditions and problems.

Non-equilibrium Properties

We are interested in obtaining averages of transport quantities which depend on the *time evolution* of the sequence of configurations of the particles in the system by means of a diffusive process. An example might be the tracer correlation factor. We will review this subject initially in general terms but later in somewhat more detail.

Again we consider a periodic box containing the particles as representing the system. We assume that the system has passed through the initial transient from an arbitrary starting configuration perhaps using the grand canonical ensemble. A *diffusive-like* transition probability is now used in the petit ensemble to specify the sequence of configurations in the irreducible Markov chain. As before, however, the distribution of configurations generated in the sequence remains Boltzmann. A statistically averaged tracer diffusion coefficient may then be determined by calculating the average of the squared displacements of the particles in time t (the Einstein equation). For an estimation of the ionic conductivity, an electric field is imposed on the system and one calculates the drift mobility in time t (eqn 3.56).

Let us now describe in a little more detail an application of the MC method for calculating tracer diffusivities and ionic conductivities. A particle is chosen at random and moved to a randomly chosen nearest neighbour site. If that neighbouring site is already occupied a particle is chosen, again at random. If the neighbouring site is vacant then the probability of moving is assessed. One introduces a barrier, E_{act}^{o}, the height of which is set equal to the maximum possible potential energy that a particle with Z-1 neighbours can have. Now, ΔE, the difference between the particular potential energy of the chosen 'diffusing' particle and the barrier height E°_{act} is calculated. ΔE will always be ≥ 0. The move is allowed with probability $\exp(-\Delta E/kT)$. A pseudo-random number, r, is generated uniform on the interval $(0,1]$. If $r < \exp(-\Delta E/kT)$ then the move is permitted. If $r > \exp(-\Delta E/kT)$ then the particle is returned to its original position. It is important to note that this diffusive transition probability, as in the PP method, depends on the state immediately *before* the particle jumps. Other transition probabilities are certainly conceivable but these have not been essayed.

To calculate a non-equilibrium property such as a tracer diffusion coefficient one must impose a tracer gradient on the system. In principle, this can be achieved by tagging even a single atom in such a way that it is distinguishable from all other atoms present. Such a tagged atom has associated with it its own tracer gradient which is maintained under steady state conditions! In practice, *every* atom is tagged, but with a *different* label. At the end of the experiment, for the calculation of the tracer diffusion coefficient, the atoms are grouped together as if they are of the same type. Direct use is then made of the Einstein (1905) equation

$$D^{*} = \frac{\langle \underline{X}^2 \rangle}{2t} \qquad (3.111)$$

where $\langle X^2 \rangle$ is the mean square displacement of the atoms in time t and unit of time is *defined* as a jump *attempt*. Use of Fick's First Law to obtain D^{*} under these somewhat bizarre tracer concentration gradients would seem to be difficult. However, if we place a formal constant tracer gradient on the system and maintain it under steady state conditions, then Fick's First Law can easily be used. Accordingly, we can tag *all* the atoms in the first plane of the array in an identical way, and, say, 90% of the atoms in the second plane,

80% in the third plane, and so on. If we maintain that gradient by
monitoring the tracer population in end planes, then D* can be
determined from

$$J^* = -D^* \frac{dC^*}{dx}$$
(3.112).

where dC*/dx is the tracer concentration gradient and J* is the
tracer flux.

In both eqns 3.111 and 3.112, D* is implied to be statistically
averaged in the sense that the jump frequency has varied from lattice
site to lattice site. It should be noted also that both the starting
configuration at t = 0 and the finishing configuration at time t
(and those in between) are characteristic of a thermal equilibrium
with respect to the *overall* distribution of particles (we make no
distinction *at this point* between tagged and untagged particles).
The calculated D* refers to that particular thermal equilibrium.

Now, D* as calculated is a product of the jump frequency and
correlation factor. Because we referred to the unit of time as a
jump attempt, the jump frequency can be further factored as the
product of the availability of vacancies to the jumping particle,
i.e., the vacancy availability factor, V and the effective jump
frequency factor, W. Since both these factors can be calculated
separately in the time average, then f can be obtained explicitly.
Because V and W are mechanical properties of the system which do
not necessarily depend on the diffusive time evolution of that
system, they may also be calculated as thermodynamic averages as
given by eqn 3.97 of section 3.5.2. This is, of course, analogous
to the fate of V and W in the PP method (see eqns 3.90 and 3.91).
In fact, it is probably obvious to the reader by now that the MC
method described here uses, in many instances, the same strategy

as the PP method but translated into the guise of a numerical simulation.

Ionic conductivity is treated in a parallel manner to that of tracer diffusion. Once thermal equilibrium has been achieved one imposes an electric field, F_x, such that the transition probability, now contains another factor, $\exp(-F_x/kT)$. For convenience, this factor is only used in the $+x$ direction. Inclusion of this factor then leads to a drift, $<X>$, downfield after time t fig. 3.25 This enables the normalized drift mobility, u, to be immediately accessed from

$$u = \frac{<X>}{F_x t}.$$
(3.113)

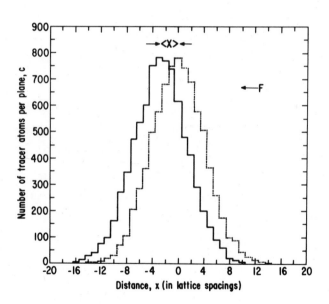

Fig. 3.25   Computer generated tracer concentration profiles in and out of an electric field; starting configuration was a δ distribution at x = 0, t = 0.

The physical correlation factor, f_I, for diffusion via the agency of vacancies, can then be found by making use of the very general Nernst-Einstein relation (Lidiard 1956)

$$\frac{u}{D_q} = \frac{q}{kT} \qquad (3.114)$$

and
$$D_q = \tfrac{1}{2} \lambda^2 \Gamma f_I \qquad (3.115)$$

where D_q is the charge-carrier diffusion coefficient (these are presumably indistinguishable entities) and q is the charge.

Although the *chemical* diffusion coefficient, $\tilde{D}$, is accessible from D* and u and the chemical potential (see eqn 3.95), it is possible to calculate it directly by starting with a lattice containing two compartments, each containing atoms at different concentrations *but each under thermal equilibrium*. The partition between the compartments is then removed and the atoms allowed to interdiffuse. $\tilde{D}$ may then be calculated using the Boltzmann-Matano analysis (Bowker and King (1978a, b)). Alternatively, one may simulate chemical diffusion under steady state conditions by maintaining two planes in the petit canonical lattice at different chemical potentials in a *grand* canonical sense, fig. 3.26. $\tilde{D}$ is then calculated from Fick's First Law

$$J = -\tilde{D} \frac{\partial C}{\partial x} \qquad (3.116)$$

where J is the net flux of atoms and $\frac{\partial C}{\partial x}$ is the compositional gradient (Murch 1980d).

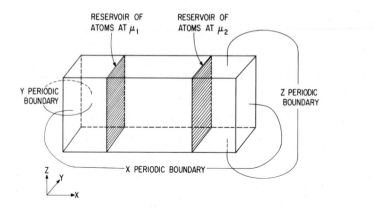

Fig. 3.26. A sketch of the realization of computer simulation of steady-state chemical diffusion.

4. <u>Conclusion</u>

In the foregoing we have reviewed the growth, over roughly a
decade, of a quantitative theory of diffusion in highly defective
solids. As a prelude to the main discussion we have traced firstly
the development of statistical mechanical treatments of nonstoichiometry. For the approach from the disordered state we may
anticipate, for the future, relatively minor theoretical refinements
and gap filling. These might include, for example, the inclusion
of next nearest neighbour interactions, a critical evaluation,
perhaps with the Ising model, of the Atlas (1970) treatment to
determine its status in an approximation hierarchy, Monte Carlo
calculations in further ionic systems, and so on. We can expect
no major progress until the problem of correlating the motional
and vibrational components of the partition function is solved.
A Monte Carlo method can conceivably be utilized in this regard
but the computational demand appears to be enormous. For the
approach from the ordered state, the description of nonstoichiometry
from the point of view of microdomains seems to be in its infancy.
Much conceptual development of this subject remains but ultimately
it would seem to be the more rewarding.

In the main discussion of the monograph we have traced the
development of the statistical mechanics of diffusion in highly
defective solids. In many respects this development has paralleled
the corresponding statistical thermodynamics of section 2, although
it has lagged behind some 10-15 years. With respect to tracer
diffusion and ionic conductivity in lattice gas systems, many gaps

still remain. These include, for example, a study of sufficient conditions for the phenomenon of physical correlation and the physical correlation factor associated with the interstitialcy mechanism. It should also be pointed out that the results so far refer to the d.c. conductivity. It has been pointed out (Sato and Kikuchi 1979) that the physical correlation factor drops from the expression for the conductivity at sufficiently high frequency and is a function of frequency up to that point. The calculation of this dependence could yield further revealing information about the nature of this factor.

The intriguing result of a tracer correlation factor greater than unity deserves further study, but not by simulating further comparably complicated mechanisms. This can only cloud the issue. What is required now is a demonstration of the effect in a simple system and which can also be handled analytically.

Ambipolar diffusion in highly defective solids has so far been handled in a manner which is too naïve. What is required now is a formal recognition of the mutual interaction among the electronic and atomic defects and a solution based upon a time-dependent statistical mechanical formalism. The long-range interactions make an analytical solution using the Path Probability method rather forbidding but a Monte Carlo simulation seems to be a good possibility.

Crystallographic shear is so far understood in only qualitative terms. Because of its importance in solid state chemistry reactions it certainly deserves a quantitative explanation. While the approach for the theorist to take is not immediately obvious, one feels that one could start by making use of the extensive simulation information available on dislocation movement in metals and extending this to ionic solids.

Results from *ab initio* defect calculations by Catlow and colleagues with the HADES program have made an important contribution to the understanding of the nature of defective solids. Ultimately, one would wish to mesh these relaxation calculations with Monte Carlo simulations. In essence, the defect formation and migration energies would then be calculated in a defect-rich system at equilibrium. In a sense, this kind of calculation is already implied in fluid-like Monte Carlo or Molecular Dynamics simulations but the huge expenditure of computation time involved makes it unlikely that these calculations will be common except in a few instances where the solid exhibits exceptionally high mobility, e.g., superionic conductivity.

While this monograph has not been specifically directed to experimental data, a few comments would seem to be appropriate. Conventional tracer diffusion and conductivity experiments have, in general, continued to furnish data in a form and a precision which are useful for testing the kinetic theory of diffusion as described in this monograph. However, there is a scarcity of tracer diffusion data of the light elements, notably oxygen. This arises, of course, because many of these elements do not have suitable radioisotopes. Recently, however, there has been a greater commitment to use the indirect *in situ* techniques: proton activation ar secondary ion mass spectrometry (SIMS). These techniques are well developed and are described, with the well-known isotope exchange method, in appendix II. We note in particular the possible use of SIMS to measure the isotope effect for oxygen diffusion. This would be a valuable and pioneering experiment.

Acknowledgments

I wish to thank my wife, Christine, for her steady encouragement and typing efforts during the preparation of this monograph. I also wish to thank my baby son, Stuart, for his many timely diversions. My thanks go also to Dr. Steve Rothman of ANL for his encouragement and inspiration and the editorial board of this monograph series for helpful and thoughtful comments.

Appendix I

Diffusion Coefficients used in the text

D^* tracer diffusion coefficient

D_q diffusion coefficient of charge carriers, defined in eqn 3.57

D_σ conductivity diffusion coefficient, defined in eqn 3.61

$\tilde{D}$ chemical diffusion coefficient, defined in eqn 3.64

D intrinsic diffusion coefficient, defined in eqn 3.64

D_{A^*} tracer diffusion coefficient, specifically of A* atoms in a host of A in the presence of vacancies V

D_I generalized diffusion coefficient of *untagged* atoms; see eqn 3.70

D_V vacancy diffusion coefficient, see eqn 3.76

Appendix II

Some Recent Experimental Techniques in Tracer Diffusion

II.1 Introductory Remarks

Many nonstoichiometric compounds are oxides and nitrides
with nonstoichiometry provided for by the respective oxygen and
nitrogen sublattices. The acquisition of tracer diffusion data
for oxygen and nitrogen[†] poses a difficult problem since neither
oxygen nor nitrogen has any long-lived radioisotopes. The
non-radioactive isotopes ^{17}O, ^{18}O and ^{15}N do exist, however,
and are readily available commercially. This has prompted
the recent development of analytical techniques applicable to
in-situ analyses of these isotopes. Such methods attempt to
replace the 'isotopic exchange method' which has been known
for many years (Auskern and Belle 1958). Since that method has
come under considerable criticism in recent years we shall briefly
review it in order to see why the recent in-situ methods are
preferable.

In the 'isotope exchange method' the sample, say an oxide,
which is usually used in the form of a polycrystalline powder
though sometimes crushed single crystals are used (Roberts et al.,
1969, Ando et al. 1976), is diffusion annealed in a closed furnace

---

[†]We might also have added fluorine but ^{18}F, with a half life of
112 mins, is just adequate as a radioactive isotope in diffusion
studies (Matzke 1971).

in, say an ^{18}O enriched atmosphere. Isotopic exchange is assumed to take place rapidly between the gas and the surface. The ^{18}O uptake of the sample, which is presumed to be *diffusion controlled*, is measured as a function of time by monitoring the ^{18}O level in the gas phase with a mass spectrometer. For evaluating the specific ^{18}O uptake of the sample corrections must be made for the ^{18}O absorption by materials other than the sample itself, for example, regions of the furnace chamber at high temperature. Such components may have an absorption area comparable to that of the sample. The evaluation and subtraction of this background may present a serious problem (Dragoo 1969). Criticism of a more general nature has been directed to (1) the fact that the method does not reveal the concentration profile of ^{18}O in the solids; the presence of short circuit paths can be neither proved nor disproved and (2) the validity of the principal assumption that the exchange rate between ^{16}O and ^{18}O at the surface is not rate determining (Simpson and Carter (1966).

Despite these shortcomings, the isotopic exchange method has frequently been used in oxygen diffusion experiments primarily because of the comparative ease of experimentation. In some cases there has been good agreement with later in-situ studies. Investigations of diffusion in highly nonstoichiometric compounds which have made use of the isotopic exchange method include Steele and Floyd (1971) and Floyd (1973) - CeO_{2-x} and CeO_2/Y_2O_3, Auskern and Belle 1958 - UO_{2+x}, Massiani et al. (1978) - Nb_2O_5, Bagshaw and Hyde (1976) - TiO_{2-x}, Sturiule and deCrescente (1965) - UN_x.

The problems associated with the isotopic exchange method
can be bypassed if the isotope is analysed in-situ. Two powerful
methods - (1) charged particle activation and (2) ion beam mass
spectrometry - have been developed for this purpose and we now
discuss these methods in turn in the following sections.

II.2   Ion Beam Mass Spectrometry

In this method one attempts to establish the concentration
profile of, say, ^{18}O in the solid by sputtering of the sample
with a primary ion beam usually of argon but krypton and nitrogen
are also sometimes used. The sputtered material thus removed
is accelerated away from the surface and monitored continuously
with a mass spectrometer.

This method has been developed in a number of diffusion studies
(Cox and Pemsler 1968, Contamin and Slodzian 1968a, b,
Marin and Contamin 1969, and Arita et al. 1979). The method is
particularly well suited to small penetrations, up to 10 μm, in
single crystal material. There are, however, difficulties in
obtaining diffusion coefficients with polycrystalline samples
because of the calibration of sputtering rate and depth. It is,
however, possible to bypass the calibration problem in the following
way. Where the penetration is very large the sample may be
sectioned *parallel* to the diffusion direction and the concentration
profile probed at discrete intervals in the diffusion direction
(Valencourt et al. 1975). A rather extensive study of oxygen diffusion
in UO_{2+x}, which has included ion beam mass spectrometry, has been
made by Contamin and Slodzian (1968), Contamin (1971), Contamin
et al. (1972). Other more recent oxygen diffusion studies using SIMS

have been made by Arita et al. (1979) in rutile, Reed and Wuensch (1980) in Al_2O_3, Perinet et al. (1980) in Cu_2O and Meyer et al. (1980) in NiO.

It is straightforward, and certainly easier than charged particle activation, to determine isotopic ratios using ion beam mass spectrometry, but, surprisingly, this has not been attempted. A probe of the ratio of ^{18}O to ^{17}O along the concentration profile gives information on the diffusion isotope effect, see section 3.2.4. This would be a most valuable experiment since the isotope effect for oxygen diffusion has never been measured in any oxide. Furthermore, the depth calibration problem referred to above does not pose a problem in the determination of the isotope effect.

II.3 Charged Particle Activation

We may identify three distinct ways in which to use charged particle nuclear reactions for probing for nonradioactive isotopes. In the first, we consider a *resonant* nuclear reaction, i.e., a reaction for which a specific energy or resonant energy of incident particles is required to induce a transition to a particular excited state. Taking $^{18}O(p,\alpha)^{15}N$ as an example, it has been found that a proton energy of 1.765 MeV gives rise to an increase in the α particle yield. The α peak arises only from ^{18}O atoms at or near the surface since the energy of the proton beam drops below this resonant value as it penetrates the solid. The reaction can then be combined with serial sectioning to determine the tracer concentration profile (Choudhury et al. 1965, Marin and Contamin 1969, Contamin et al. 1972).

A variation of this first method is the use of a reaction such as $^{18}O(p,n)^{18}F$. Although this resonant reaction can be used in conjunction with neutron counting (Murch 1973), it is more usual to take advantage of the product nucleus, ^{18}F, which has a half-life of 112 minutes and decays by positron emission, and use an autoradiographic procedure. In specific tracer diffusion studies (Holt 1967, Anderson 1967, Bradhurst and de Bruin 1968, Murch et al. (1975), the sample was bevelled at a small angle to 'expand' the tracer concentration profile. Following proton irradiation of about an hour and a short period thereafter to allow short-lived radioactivity to decay, an autoradiograph is taken which is subsequently scanned with a microdensitometer. A nonresonant reaction such as ^{17}O(d,n)^{18}F may also be used since the penetration range of deuterons in most materials is quite small and is comparable to that of the 'surface' region probed with the reaction: ^{18}O(p,n)^{18}F.

In the second way of applying nuclear reactions, one starts with a resonant reaction such as ^{18}O(p,γ)^{19}F, in particular a reaction in which the emitted hard γ radiation is not stopped substantially by the matrix. If the proton energy is raised above the resonance, the average proton energy will drop to the resonance value, in this case 1.167 MeV (Zelenskii et al. 1970) in some region below the surface and excite ^{18}O atoms in that region. Clearly then, by using progressively higher incident proton energies above that resonance, one can probe successively deeper layers. The limit of depth analysis performed in this way depends on the stopping power of the material and the existence of higher resonances. Commonly, the maximum depth which can be probed is of the order of 10 μm, a depth which is rather small

for a tracer diffusion experiment. This method of analysis
has been used by Hadari et al. (1971), Derry et al. (1971),
Ollerhead et al. (1966), and Cox and Roy (1966).

For the third method, one takes advantage of the following
information. For a reaction like $^{18}O(p,\alpha)^{15}N$, Palmer (1965)
has shown that within the precision given by energy straggling
of the particles, the measured α particle energy determines the
depth at which the α particle is formed. The height of the α
spectrum at this energy depends on the product of the ^{18}O
concentration at that depth and the reaction cross section for
the proton energy at the same depth. The shape of the α spectrum
is intimately related to the depth distribution of the ^{18}O in the
solid and hence the tracer diffusion coefficient. This distribution can be determined either by comparison with a sample of
homogeneously distributed ^{18}O or by using cross section data
when energy losses of both protons and α particles are known.
A *single* incident proton energy, which is chosen well above any
resonance so that the variation of reaction cross section is
smooth, can thus result in an estimate of the tracer diffusion
coefficient. This method has been used by Robin (1971) and
Amsel et al. (1968).

References

Adda, Y., and Philibert, J., 1966, *La Diffusion dans les Solides*

    (Paris: Presses Universitaires de Paris)

Alex, K., and McLellan, R. B., 1971, *Acta Metall.*, $\underline{19}$, 439.

Allen, R. L., and Moore, W. J., 1959, *J. Phys. Chem. Solids*, $\underline{63}$, 223.

Allnatt, A. R., and Cohen, M. H., 1964, *J. Chem. Phys.*, $\underline{40}$, 1860, 1871.

Amsel, G., Béranger, G., de Gelas, B., and La Combe, P., 1968,

    J. Appl. Phys., $\underline{39}$, 2246.

Amsel, G., and Samuel, D., 1967, *Analyt. Chem.*, $\underline{39}$, 1689.

Anderson, J. S., 1970, *Problems of Nonstoichiometry*, edited by

    A. Rabenau (Amsterdam: North Holland).

Anderson, J. S., 1969, *Bull. Soc. Chim. France*, $\underline{7}$, 2203.

Anderson, J. S., 1946, *Proc. Roy. Soc. Ser A.*, $\underline{185}$, 69.

Anderson, J. S., 1972, *Surface and Defect Properties of Solids*, Vol. 1

    (London: The Chemical Society).

Anderson, J. S., and Tilley, R. J. D, 1974, *Surface and Defect Properties*

    of Solids, Vol. 3 (London: The Chemical Society).

Anderson, J. S., and Hyde, B. G., 1967, *J. Phys. Chem. Solids*, $\underline{28}$, 1393.

Anderson, J. S., and Tilley, R. J. D., 1970, *J. Sol. St. Chem.*, $\underline{2}$, 472.

Anderson, R. L., 1967, Ph.D. Thesis M.I.T.

Andersson, S., and Wadsley, A. D., 1966, *Nature*, $\underline{211}$, 581.

Ando, K., Oishi, Y., and Hidaka, Y., 1976, *J. Chem. Phys.*, $\underline{65}$, 2751.

Arita, M., Hosoya, M., Kobayashi, M., and Someno, M., 1979,

    J. Am. Ceram. Soc., $\underline{62}$, 443.

Ariya, S. M., and Morozova, M. P., 1958, *J. Gen. Chem.*, (USSR) $\underline{28}$, 2617.

Ariya, S. M., and Popov, Y. G., 1962, *J. Gen. Chem.*, (USSR), $\underline{32}$, 2077.

Arsenault, R. S., Beeler, J. R., and Simmons, J. A., 1976, *International*

    Conference of Computer Simulation for Materials Applications,

    published in *Nucl. Metall.*, Vol. 20.

Atlas, L. M., 1968a, *J. Phys. Chem. Solids*, 29, 91.

Atlas, L. M., 1968b, *J. Phys. Chem. Solids*, 29, 1349.

Atlas, L. M., 1970, *Chemistry of Extended Defects in Non-Metallic Soli*
edited by L. Eyring and M. O'Keeffe (Amsterdam: North Holland)

Auskern, A. B., and Belle, J., 1958, *J. Nucl. Mater.*, 3, 267.

Bagshaw, A. N., and Hyde, B. G., 1976, *J. Phys. Chem. Solids*, 37, 835.

Baker, B. G., 1966, *J. Chem. Phys.*, 45, 2694.

Bardeen, J., and Herring, C., 1952, *Imperfections in Nearly Perfect
Crystals*, edited by W. Shockley (New York: Wiley).

Barker, W. W., and Knop, O., 1971, *Proc. Brit. Ceram. Soc.*, 19, 15.

Barsis, E., and Taylor, A., 1966, *J. Chem. Phys.*, 45, 1154.

Bartkowicz, I., and Mrowec, S., 1972, *Phys. Stat. Sol.*, (b) 49, 101.

Belle, J., 1969, *J. Nucl. Mater.*, 30, 3.

Bennett, C. H., 1975, *Diffusion in Solids: Recent Developments*, edite
by A. S. Nowick and J. J. Burton (New York: Academic).

Benoist, P., Bocquet, J., and Lafore, P., 1977, *Acta Metall.*, 25, 26!

Bentle, G. G., 1968, *J. Appl. Phys.*, 39, 4036.

Bevan, D. J. M., and Kordis, I., 1964, *J. Inorg. Nucl. Chem.*, 26, 1509

Binder, K., 1979, Topics in Current Physics, Vol. 7 (New York:
Springer-Verlag).

Bird, J. R., Russell, L. H., and Murch, G. E., 1974, 5th AINSE Nucl.
Phys. Conference, Canberra.

Boureau, G., and Campserveux, J., 1977, *Phil. Mag.*, 36, 9.

Bowker, M., and King, D. A., 1978a, *Surface Sc.*, 71, 583.

Bowker, M., and King, D. A., 1978b, *Surface Sc.*, 72, 208.

Bradhurst, D. H., and de Bruin, H. J., 1969, *J. Aust. Ceram. Soc.*, 5, 2

Brebrick, R. F., 1958, *J. Phys. Chem. Solids*, 4, 190.

Bursill, L. A., and Hyde, B. G., 1971, *Phil. Mag.*, 23, 3.

Bursill, L. A., and Hyde, B. G., 1972, *Prog. Sol. St. Chem.*, 7, 1977.

Bustard, L. D., 1979, Ph.D. Thesis, Cornell University.

Carter, R. E., and Roth, W. L., 1968, *Electromotive Force Measurements in High Temperature Systems*, (London: Institute of Mining and Metallurgy).

Catlow, C. R. A., Lidiard, A. B., and Norgett, M. J., 1975, *J. Phys. C.*, 8, L435.

Catlow, C. R. A., 1976, *J. Nucl. Mater.*, 60, 151.

Catlow, C. R. A., and Fender, B. E. F., 1975, *J. Phys. C.*, 8, 3267.

Chen, W. K., Petersen, N. L., and Reeves, W. T., 1969, *Phys. Rev.*, 186, 887.

Chen, W. K., and Petersen, N. L., 1975a, *Mass Transport Phenomena in Ceramics*, edited by A. R. Cooper and A. H. Heuer (New York: Plenum).

Chen, W. K., and Petersen, N. L., 1975b, *J. Phys. Chem. Solids*, 36, 1097.

Chesnut, D. A., and Salsburg, Z. W., 1963, *J. Chem. Phys.*, 38, 2861.

Choudhury, A., Palmer, D. W., Amsel, G., Carien, H., and Baruch, P., 1965, *Sol. St. Commun*, 4, 3107.

Contamin, P., 1971, D. Eng. Thesis, University of Grenoble.

Contamin, P., Bacmann, J. J., and Marin, J. F., 1972, *J. Nucl. Mater.*, 42, 54.

Contamin, P., and Slodzian, G., 1968, *Compt. Rend. (Acad. Sc. Paris).*, *Sec. C.*, 267, 805.

Corish, J., and Jacobs, P. W. M., 1972, *J. Phys. Chem. Solids*, 33, 1799.

Cox, B., and Roy, C., 1966, *Electrochem. Tech.*, 4, 121.

DeBruin, H. J., and Murch, G. E., 1973, *Phil. Mag.*, 27, 1475.

Derry, D. J., Lees, D. G., and Calvert, J. M., 1971, *Proc. Brit. Ceram. Soc.*, 19, 77.

Douglas, T. B., 1964, *J. Chem. Phys.*, 40, 2248.

Dragoo, A. L., 1968, *J. Res. (NBS)*, 72A, 157.

Einstein, A., 1905, *Ann. Phys.*, 17, 549.

Eyring, H., and Marchi, R. P., 1963, *J. Chem. Educ.*, 40, 562.

Eyring, L., and O'Keeffe, M., 1970, *The Chemistry of Extended Defects Non-Metallic Solids*, (Amsterdam: North Holland).

Fedders, P. A., and Sankey, O. F., 1978, *Phys. Rev.*, B18, 5938.

Floyd, J. M., 1973, *Ind. J. Technol.*, 11, 589.

Foster, J. S., and Dooley, D. W., 1977, *Acta Metall.*, 25, 759.

Funke, K., 1976, *Prog. Sol. St. Chem.*, 11, 345.

Gallagher, P. T., Lambert, J. A., Oates, W. A., 1969, *Trans AIME*, 245,

Gillan, M. J., and Dixon, M., 1980, *J. phys. C.*, 10, 1901.

Greenwood, R., and Howe, A. T., 1972, *J. Chem. Soc.*, (Dalton), 1, 110.

Gschwend, K., Sato, H., and Kikuchi, R., 1977, *Bull. Am. Phys. Soc.*, 22, 442.

Hadari, Z., Kroupp, M., and Wolfson, Y., 1971, *J. Appl. Phys.*, 42, 534

Hagemark, K., 1963, Kjeller Report, KR-48.

Hagenmuller, P., and van Gool, W., 1978, *Solid Electrolytes*, (New York: Academic).

Haven, Y., 1978, *Solid Electrolytes*, edited by P. Hagenmuller and W. van Gool (New York: Academic).

Hayes, W., 1978, *Contemp. Phys.*, 19, 469.

Heyne, L., 1968, *Mass Transport in Oxides*, edited by J. B. Wachtman and A. D. Franklin, NBS special publication no 296.

Hill, T. L., 1960, An Introduction to Statistical Thermodynamics (Reading, Mass.: Addison-Wesley).

Holt, J. B., 1967, *Proc. Brit. Ceram. Soc.*, 9, 157.

Holt, J. B., and Almassy, M. Y., 1969, *J. Am. Ceram. Soc.*, 52, 631.

Hooper, A., 1978, *Contemp. Phys.*, 19, 147.

Howard, R. E., and Lidiard, A. B., 1964, *Rept. Prog. Phys.*, 27, 161.

I.A.E.A. panel, 1965, *Tech. Rept. Ser.*, *Vo. 34*, (Vienna: IAEA).

Ingram, M. D., and Vincent, C. A., 1977, *Ann. Rept. Prog. Chem.*, 74, 23.

Jordan, P., and Pochon, M., 1957, *Melv. Phys. Acta.*, 30, 33.

Kikuchi, 12, 1951, *Phys. Rev.*, 81, 788.

Kikuchi, R., and Sato, H., 1974, *Acta Metall.*, 22, 1099.

Kim, K. K., Mundy, J. N., and Chen, W. K., 1979, *J. Phys. Chem. Solids*,
    40, 743.

Kirkwood, J. G., 1935, *J. Chem. Phys.*, 3, 300.

Klauber, C., 1977, Honours Thesis, Flinders University of South Aust.

Koch, F. B., and Cohen, J. B., 1969, *Acta Cryst.*, B25, 275.

Kofstad, P., 1972, *Nonstoichiometry, Diffusion and Electrical Conductivity
    in Binary Metal Oxides*, (New York: Wiley).

Krivoglaz, M. A., Smirnov, A. A., 1964, *Theory of Order/Disorder in
    Alloys*, (New York: Elsevier).

Kröger, F. A., 1964, *Chemistry of Imperfect Crystals*, (Amsterdam:
    North Holland).

Kvist, A., and Tärneberg, R., 1970, *Z. Naturforsch.*, 25a, 257.

LeClaire, A. D., 1970, *Physical Chemistry - An Advanced Treatise*, Vol. 10,
    edited by H. Eyring, D. Henderson and W. Jost (New York: Academic).

LeClaire, A. D., 1975, *Mass Transport Phenomena in Ceramics*, edited by
    A. R. Cooper and A. H. Heuer (New York: Plenum).

Lee, H. M., 1974, *Metall. Trans.*, 5, 787.

Libowitz, G. G., 1968, *Mass Transport in Oxides*, edited by J. B. Wachtman
    and A. D. Franklin, NBS special publication no. 296.

Lidiard, A. B., 1955, *Phil. Mag.*, 46, 1218.

Lidiard, A. B., 1966, *J. Nucl. Mater.*, 19, 106.

Lidiard, A. B., 1956, *Handbuch der Physik*, 20, 246 (Berlin: Springer-Verlag).

McGeehin, P., and Hooper, A., 1977, *J. Mater. Sc.*, 12, 1.

McLellan, R. B., Gerrard, T. L., Horowitz, S. J., and Sprague, J. A.,
1967, *Trans. TMS-AIME*, 234, 528.

McQuillan, A. D., 1967, *Phase Stability in Metals and Alloys*, edited b
P. S. Rudman, J. Stringer and R. J. Jaffee (New York: McGraw Hil

Magnéli, A., 1953, *Acta Cryst.*, 6, 495.

Mahan, G. D., and Roth, W. L., 1976, *Superionic Conductors*,
(New York: Plenum).

Manes, L., and Manes-Pozzi, B. M. 1976, *Plutonium and other Actinides*,
edited by R. Lindner and M. Blank (Amsterdam: North-Holland).

Manes, L., Sørensen, O. T., Mari, C., and Ray, I., 1979, *Thermodynamic
of Nuclear Materials* (Vienna: IAEA).

Manning, J. R., 1968, *Diffusion Kinetics for Atoms in Crystals*,
(Princeton, N.J.:   Van Nostrand).

Marin, J. F., and Contamin, D., 1969, *J. Nucl. Mater.*, 30, 16.

Martin, S. L. H., and Rees, A. L. G., *Trans. Farad. Soc.*, 50, 343.

Massiani, Y., Crousier, J. P., and Streiff, R., 1978,   *J. Sol. St. Che*
23, 415.

Matzke, Hj., 1966, Tech. Rept., AECL-2585.

Matzke, Hj., 1971, *J. Chem. Phys.*, 32, 437.

Metropolis, N. A., Rosenbluth, A. W. Rosenbluth, M. N., Teller, A. H.,
and Teller, E., 1953, *J. Chem. Phys.*, 21, 1087.

Merritt, R. R., 1970, Ph. D. Thesis, University of Western Australia.

Meyer, M., Barzebat, S., El Houch, C., and Talon, R., 1980, Third
Europhysical Conference on Lattice Defects in Ionic Crystals,
Canterbury, 1979, to be published in J. de Physique (Colloques).

Mitra, S. K., and Allnatt, A. R., 1979, *J. Phys. C.*, 12, 2261.

Moon, K. A., 1963, *Trans.   TMS-AIME*, 227, 1116

Mott, N. F., and Gurney, R. W., 1950, *Electronic Processes in Ionic
Crystals*, (Oxford: Claredon).

Murch, G. E., 1973, Ph.D. Thesis, Flinders University of South Aust.

Murch, G. E., 1975a, *J. Nucl. Mater.*, 57, 239.

Murch, G. E., 1975b, *Phil. Mag.*, 32, 1129.

Murch, G. E., 1979, *Acta Metall.*, 27, 1701.

Murch, G. E., 1980a, *Phil. Mag.*, 41A, 157.

Murch, G. E., 1980b, to be published.

Murch, G. E., 1980c, *Phil. Mag.*, in the press.

Murch, G. E., 1980d, *J. Phys. Chem. Solids*, to be published.

Murch, G. E., 1980e, *Phil. Mag.*, to be published.

Murch, G. E., and Rothman, S. J., 1980, *Phil. Mag.*, in the press.

Murch, G. E., and Nowick, A. S., 1980, to be published.

Murch, G. E., Bradhurst, D. H., and de Bruin, N. J., 1975, *Phil. Mag.*, 32, 1141.

Murch, G. E., and Thorn, R. J., 1976, *Phil. Mag.*, 34, 299.

Murch, G. E., and Thorn, R. J., 1977a, *Phil. Mag.*, 35, 493.

Murch, G. E., and Thorn, R. J., 1977b, *Phil. Mag.*, 35, 1441.

Murch, G. E., and Thorn, R. J., 1977c, *J. Phys. Chem. Solids*, 38, 789.

Murch, G. E., and Thorn, R. J., 1977d, *Phil. Mag.*, 36, 517.

Murch, G. E., and Thorn, R. J., 1977e, *Phil. Mag.*, 36, 529.

Murch, G. E., and Thorn, R. J., 1978a, *J. Nucl. Mater.*, 71, 219.

Murch, G. E., and Thorn, R. J., 1978b, *Phil. Mag.*, 37, 85.

Murch, G. E., and Thorn, R. J., 1978c, Electrochem. Soc. Spring Meeting, Seattle.

Murch, G. E., and Thorn, R. J., 1978d, *J. Comput. Phys.*, 29, 237.

Murch, G. E., and Thorn, R. J., 1979a, *Phil. Mag.*, 39A, 673.

Murch, G. E., and Thorn, R. J., 1979b, *J. Phys. Chem. Solids*, 40, 389.

Murch, G. E., and Thorn, R. J., 1979c, *Acta Metall.*, 27, 201.

Murch, G. E., and Thorn, R. J., 1979d, *Phil. Mag.*, 40A, 477.

Murch, G. E., and Thorn, R. J., 1979e, *J. Nucl. Mater.*, 82, 430.

Murch, G. E., and Thorn, R. J., 1979f, *Phil. Mag.*, 39A, 259.

Norris, D. I. R., 1977, *J. Nucl. Mater.*, 68, 13.

Oates, W. A., Lambert, J. A., and Gallagher, P. T., 1969, *Trans. TMS-AIME*, 245, 47.

O'Keeffe, M., 1970, *Chemistry of Extended Defects in Non-Metallic Soli* edited by L. Eyring and M. O'Keeffe (Amsterdam; North Holland).

O'Keeffe, M., 1971, *Proc. Brit. Ceram. Soc.*, 19, 1.

O'Keeffe, M. 1976, *Superionic Conductors*, edited by G. D. Mahan and W. L. Roth (New York: Plenum).

O'Keeffe, M., and Hyde, B. G., 1976, *Phil. Mag.*, 33, 219.

Ollerhead, R. W., Alnquist, E., and Kuehner, J. A., 1966, *J. Appl. Phy* 37, 2440.

Owens, B. B., and Argue, G. R., 1967, *Science*, 157, 308.

Palmer, D. W., 1965, *Nucl. Inst. Methods*, 38, 187.

Perinet, F., Barzebat, S., and Monty, C., 1980, Third Europhysical Conference on Lattice Defects in Ionic Crystals, Canterbury, 1979, to be published in J. de Physique (Colloques).

Perron, P. O., 1968, Tech. Rept., AECL-3072.

Rahman, A., 1976, *J. Chem. Phys.*, 65, 4845.

Reed, D., and Wuensch, B., 1980, *J. Am. Ceram. Soc.*, 63, 88.

Rees, A. L. G., 1954, *Trans. Farad. Soc.*, 50, 335.

Roberts, L. E. J., and Markin, T. L., 1967, *Proc. Brit. Ceram. Soc.*, 18, 201.

Roberts, L. E. J., Wheeler, V. J., and Perrin, A., unpublished work cit by Belle (1969).

Robin, R. C., 1971, Ph.D. Thesis, University Microfilms 71-22843.

Roth, W. L., Reidinger, F., and LaPlaca, S., 1976, *Superionic Conductor* edited by G. D. Mahan and W. L. Roth (New York: Plenum).

Russell, L. H., and Murch, G. E., 1973, Royal Aust. Chem. Inst. Symposi on Analyt. Chem. Sydney.

Salamon, M. B., 1979, *Physics of Superionic Conductors*, (Berlin: Springer-Verlag).

Sato, H., and Gschwend, K., 1980, to be published.

Sato, H., and Kikuchi, R., 1969, *J. Chem. Phys.*, 51, 161.

Sato, H., and Kikuchi, R., 1970, *J. Chem. Phys.*, 53, 2702.

Sato, H., and Kikuchi, R., 1971a, *J. Chem. Phys.*, 55, 677.

Sato, H., and Kikuchi, R., 1971b, *J. Chem. Phys.*, 55, 702.

Sato, H., and Kikuchi, R., 1975, *Mass Transport in Ceramics*, edited by

    A. R. Cooper and A. H. Heuer (New York: Plenum).

Sato, H., and Kikuchi, R., 1976, *Superionic Conductors*,

    G. D. Mahan and W. L. Roth (New York: Plenum).

Sato, H., and Kikuchi, R., 1977, *J. de Physique*, 38, C7-159.

Sato, H., and Kikuchi, R., 1979, *Fast Ion Transport in Solids*, edited by

    P. Vashishta, J. N. Mundy and G. K. Shenoy (Amsterdam: North Holland).

Schoen, A., 1958, *Phys. Rev. Lett.*, 1, 138.

Smeltzer, W. W., and Young, D. J., (1976), *Prog. Sol. St. Chem.*, 10, 17.

Shahi, K., 1977, *Phys. Stat. Sol.*, (a) 41, 11.

Shewmon, P. G., 1963, *Diffusion in Solids*, (New York: McGraw-Hill).

Schottky, W., and Wagner, C., 1930, *Z. Physical Chem.*, B11, 163.

Shockley, W. 1938, *J. Chem. Phys.*, 6, 130.

Simpson, L. A., and Carter, R. E., *J. Amer. Ceram. Soc.*, 49, 139.

Speiser, R., and Spretnak, J. W., 1955, *Trans. Am. Soc. Metals.*, 47, 493.

Steele, B. C. H., 1972, *Solid State Chemistry*, Vol. 10 edited by

    L. E. J. Roberts (London: Butterworths).

Steele, B. C. H., 1973, *Fast Ion Transport in Solids*, edited by

    W. van Gool (Amsterdam: North Holland).

Steele, B. C. H., and Floyd, J. M., 1971, *Proc. Brit. Ceram. Soc.*, 19, 55.

Stripp, K. F., and Kirkwood, J. G., 1954, *J. Chem. Phys.*, 22, 1579.

Sturiale, T. J., and de Crescente, M. A., 1965, Tech. Rept. PWAC-477.

Subbaswamy, K. R., and Mahan, G. D., 1976, *Phys. Rev. Lett.*, 37, 642.

Tateno, J., 1979, *J. Solid. St. Chem.*, 23, 163.

Tetenbaum, M., and Hunt, P. D., 1971, *J. Nucl. Mater.*, 40, 104.

Tharmalingham, K., and Lidiard, A. B., 1959, *Phil. Mag.*, 4, 899.

Tilley, R. J. D., 1972, *Solid State Chemistry*, *Vol. 10*, edited by
    L. E. J. Roberts (London: Butterworths).

Thorn, R. J., 1970, *Chemistry of Extended Defects in Non-Metallic Solid*
    edited by L. Eyring and M. O'Keeffe (Amsterdam: North Holland).

Thorn, R. J., and Winslow, G. H., 1967, *Advances in High Temperature*
    Chemistry, *Vol. 1*, edited by L. Eyring (New York: Academic).

Thorn, R. J., and Winslow, G. H., 1966a, *J. Chem. Phys.*, 44, 2632.

Thorn, R. J., and Winslow, G. H., 1966b, *Thermodynamics with Emphasis*
    Nuclear Materials and Atomic Transport in Solids, *Vol. 2.*,
    (Vienna: IAEA).

Thorn, R. J., and Winslow, G. H., 1966c, *J. Chem. Phys.*, 44, 2822.

Thornber, M. R., and Bevan, D. J. M., 1970, *J. Sol. St. Chem.*, 1,
    526, 545.

Ubbelohde, A. R., 1957, *Q. Rev. Chem. Soc. London*, 11, 246.

Ubbelohde, A. R., 1966, *J. Chim. Phys. Chim. Biol.*, 62, 33.

Valleau, J. P., and Torrie, G. M., 1976, *Mod. Theor. Chem.* Vol. 5,
    edited by B. J. Berne (New York: Plenum).

Valleau, J. P., and Whittingham, S. G., 1976, *Mod. Theor. Chem.*, Vol. 5
    edited by B. J. Berne (New York: Plenum).

Valencourt, L. R., Johnson, C. E., Steidl, D. V., and Davis, H. T., 197
    J. Nucl. Mater., 58, 293.

Van Gool, W., 1973, *Fast Ion Transport in Solids*, (Amsterdam: North
    Holland).

Vashishta, P., Mundy, J. N., and Shenoy, G. K., 1979, *Fast Ion Transport in Solids*, (Amsterdam: North Holland).

Vineyard, G. H., 1957, *J. Chem. Phys.*, 3, 121.

Volpe, M. L., Petersen, N. L., and Reddy, J., 1971, *Phys. Rev.*, B3. 141.

Wadsley, A. D., 1963, *Nonstoichiometric Compounds*, edited by L. Mandelcorn (New York: Academic).

Wagner, C., 1933, *Z. Physik. Chem.*, B21, 25.

Wagner, C., 1976, *Prog. Sol. St. Chem.*, 10, 1.

Weber, M. D., and Friauf, R. J., 1969, *J. Phys. Chem. Solids*, 30, 407.

Whittingham, M. S., and Huggins, R. A., 1971a, *J. Chem. Phys.*, 54, 414.

Whittingham, M. S., and Huggins, R. A., 1971b, *J. Electrochem. Soc.*, 118, 1.

Widom, B., 1963, *J. Chem. Phys.*, 39, 2808.

Willis, B. T. M., 1964, *Proc. Brit. Ceram. Soc.*, 1, 9.

Wolf, D., 1979, *J. Phys. Chem. Solids*, 40, 757.

Wood, W. W., and Erpenbeck, J. J., 1976, *Ann. Rev. Chem.*, 27, 319.

Yao, T. F. Y., and Kummer, J. T., 1967, *J. Inorg. Nucl. Chem.*, 29, 2453.

Yureck, G., 1978, Electrochemical Society Spring Meeting, Seattle.

Zelenskii, V. F., Kharskov, O. N., Kulakov, U. S., and Skakun, N. A., 1970, *Zashchita Mettalov.*, 6, 246.